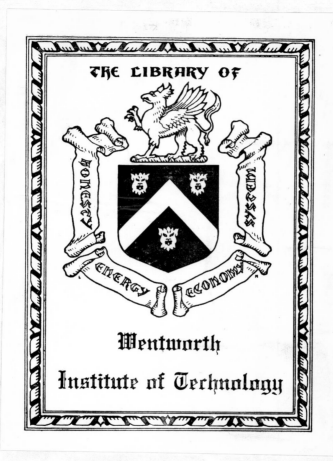

Fundamentals of Survey Measurement and Analysis

ASPECTS OF MODERN LAND SURVEYING

Series editor: J. R. SMITH, A.R.I.C.S.

OPTICAL DISTANCE MEASUREMENT
by J. R. Smith, A.R.I.C.S.

MODERN THEODOLITES AND LEVELS
by M. A. R. Cooper, B.Sc., A.R.I.C.S.

ELECTROMAGNETIC DISTANCE MEASUREMENT
by C. D. Burnside, M.B.E., B.Sc., A.R.I.C.S.

DESK CALCULATORS
by J. R. Smith, A.R.I.C.S.

FUNDAMENTALS OF SURVEY MEASUREMENT AND ANALYSIS
by M. A. R. Cooper, B.Sc., A.R.I.C.S.

HYDROGRAPHY FOR THE SURVEYOR AND ENGINEER
by Lt.-Cdr. A. E. Ingham, R.N.(Retd), A.R.I.C.S.

# Fundamentals of survey measurement and analysis

M. A. R. COOPER, B.Sc., A.R.I.C.S.
Department of Civil Engineering
The City University, London

Crosby Lockwood Staples   London

Granada Publishing Limited
First published in Great Britain 1974 by
Crosby Lockwood Staples
Frogmore St Albans Herts and 3 Upper James Street London W1R 4BP

ISBN 0 258 96871 0

Printed in Great Britain by
William Clowes & Sons, Limited
London, Beccles and Colchester

# Preface

The advances in electronics which have been made during the last fifteen years have resulted in changes in two major land surveying procedures: the measurement of distance and computation. Some aspects of developments in these areas are described and analysed in two volumes of this series (Burnside's *Electromagnetic Distance Measurement* and Smith's *Desk Calculators*). It is probable that, in the next few years, further developments will take place (particularly in the automatic recording and computation of field data), and the land surveyor and engineering surveyor will be faced by a wide variety of procedures to fulfil a specific task. The choice of an appropriate procedure depends very much upon an adequate understanding of the fundamentals of survey measurement and analysis, and this book has been written to help the reader achieve such an understanding.

Land surveying procedures entail the application of physical and mathematical principles. The mathematics included in the book is not completely rigorous, although more nearly so than is usual in land survey texts, and will be adequate for most survey practitioners. Where any reader is dissatisfied with the rigour of a derivation, the references given will provide a fuller explanation. A working knowledge of elementary calculus and matrix algebra is assumed, but a whole chapter (Chapter 2) is devoted to some statistics of a random variable, since modern analysis procedures depend to a large extent upon such principles.

The book is written primarily for students studying for degrees in land surveying and civil engineering, but it is also hoped that it will be of assistance to surveyors and engineers in practice, and particularly to those whose formal education preceded the developments in measurement and analysis explained here.

*M. A. R. Cooper*
1973

I wish to express my gratitude to Alfred Kenny, who drew the diagrams. I am also grateful to F. J. Leahy of the University of Melbourne, N. E. Lindsey of The City University and J. R. Smith of Portsmouth Polytechnic, all of whom read the text at an early stage and pointed out some mistakes and omissions, and whose suggestions have been very valuable. My thanks go to my wife for her painstaking interpretation and correction of the manuscript. In spite of this assistance, any mistakes which remain must be attributed either to my own carelessness or obduracy.

# Contents

Contents

# CHAPTER ONE

# The Measurement Process

## 1.1 Definition of Measurement, Accuracy and Error

The measurement of a quantity means the assignation of a numerical value to represent that quantity. Land and engineering surveys are concerned with the measurement of lengths and angles, and the first stage in the measurement process is a definition of the *units* of measurement. Secondly, a *mathematical model* is constructed to simplify the physical reality of the components of measurement and, thirdly, a *procedure* for the measurement is prescribed and then followed. As a result of following the procedure, *observations* or *readings* are obtained which are then transformed in accordance with the mathematical model to give the *values* required. These concepts are examined in more detail in the next four sections.

The *error* (*e*) of a measurement is defined as

$$e = M - T \qquad (1.1)$$

where $M$ is the measured value and $T$ is the true value. The *accuracy* of a measurement is the nearness of $M$ to $T$, so a small error means a high accuracy and *vice versa*. Unfortunately, except in a few idealised cases the true value ($T$) of a quantity measured in surveying is unknown, so the error and accuracy of the measurement are also unknown. The two following examples illustrate the indeterminate nature of accuracy.

If a line of levels starts on a bench mark and closes on the same bench mark, then the true difference in height around the circuit is zero, provided that the time taken to observe the staff readings around the circuit is itself zero. If a finite time is taken, then one has to admit the possibility of the bench mark moving with a vertical component during the levelling. Although this possibility may often be very slight, it prevents the true difference in height being known. Similarly, the sum of the angles in a plane triangle is 180° exactly. But if a theodolite is set up in turn at each of three ground points (the direction of gravity being constant over the area of the triangle), and the angle between the other two points measured, then the true value of the sum of the three angles measured is 180° exactly, provided, once again, that the time taken to carry out the measurement procedure is zero.

In the rare cases where the true value of a quantity is known, the finite time necessary to measure that quantity means that one cannot be certain that the quantity has remained constant during the measurement. Of course, in most survey work any movement of 'permanent' marks is likely to be small, but not certain to be zero.

Since accuracy is indeterminate in practice, we have to resort to other concepts if we are to assess the reliability of a measurement or decide which of two or more measurements is 'best' (having defined what we mean by 'best'). The *precision* of a measurement enables us to make such judgements. Precision is defined in statistical terms in section 2.5, and Chapter 3 deals in detail with this aspect of measurement. Precision thus defined is an indication of the spread of measured values of a quantity. Several values grouped closely together are said to constitute a more precise set of measurements than one with a broader spread of values.

'Precise' has another, general, meaning: manufacturers of surveying equipment refer to 'precise levels', drawing instruments can be 'precision instruments' and a watch may have a 'precision movement'. Used in this context, 'precise' denotes something made to a higher manufacturing standard, and having more refinements than usual. For example, a watch with a seconds hand can be said to be more precise than one without; it could well be less accurate in spite of this. Neither this use of 'precise' nor the specific definition of precision in section 2.5 should be confused with accuracy.

Although absolute error is indeterminate, sources of error can be identified, the errors classified and steps taken to reduce them. This is best done by considering the measurement process in more detail.

## 1.2 Units of Measurement

The units used in this book are based on the International System (SI) which was accepted by the Eleventh General Conference of Weights and Measures (CGPM) in Paris in 1960. The SI units themselves are based on the metric system.

The maintenance of standards in the UK is the responsibility of the National Physical Laboratory (NPL) and [2] gives information necessary for the conversion from the yard–pound–second system to SI units.

### 1.2.1 *The SI Units of Length, Mass and Time*
The unit of length is the *metre*, which is defined as a multiple of the wavelength of the radiated electromagnetic energy resulting from a specified atomic transition. The radiation is that produced by a transition in between the energy levels $2p_{10}$ and $5d_5$ of the krypton–86 atom, and the metre is 1 650 763·73 times the wavelength of that radiation in a vacuum. (XI CGPM, 1960)

The unit of mass is the *kilogramme* and is equal to the mass of the international prototype kilogramme kept by the Bureau International des Poids et Mesures at Sèvres. (III CGPM, 1901)

The unit of time is the *second* which is defined as the time interval occupied by a number of cycles of the radiated electromagnetic energy resulting from a specified atomic transition. The radiation is that produced by the transition $(F = 4, M_F = 0)$–$(F = 3, M_F = 0)$ of the caesium–133 atom and the second is the time interval occupied by 9 192 631 770 cycles of that radiation. (XII CGPM, 1964)

Although these definitions of length and time can be considered as exact, in practice their usefulness depends upon the repeatability of the specific atomic transition under different conditions in different standards laboratories. It is thought that the ultimate accuracy with which the krypton transition can be repeated is of the order of 1 in $10^9$, although 1 in $10^8$ is more usual. It may be possible in the future to make use of a servo-controlled gas laser emission as a definition of length, which will have repeatability of the order of 1 in $10^{11}$. The practical application of the caesium transition defining the second is repeatable to better than 1 in $10^{10}$.

The velocity of light in a vacuum (the 1972 value is $c = 299\ 792·5 \pm 0·4$ km/s) is known only to about 1 in $10^7$ so, in spite of time measurement of the order of $10^{-9}$, distances can theoretically be deduced only to 1 in $10^7$. Thus the greatest (theoretical) uncertainty in electromagnetic distance measurement is the velocity of the signal. A better determination of the velocity of light is desirable for another reason: it is the square root of the factor which is used to convert electrostatic units to electromagnetic units. The most precise method for measuring capacitance gives a value in electrostatic units which has to be converted to a value in farads, the SI electromagnetic units. Thus the accuracy of the assignment of values to electric standards depends upon the accuracy of the value for the velocity of light in a vacuum.

### 1.2.2 *The Units of Plane Angular Measure*

The SI unit is the *radian* (rad) which is the smaller angle between two radii of a circle which cut off on the circumference an arc equal in length to the radius. There are therefore $2\pi$ radians in a full circle. Since $2\pi$ is an irrational number, the use of the radian as a practical unit is inconvenient. There are two main systems of practical units: the *sexagesimal* system and the *centesimal* system. The former is in common use and has as its basic unit the *degree*, defined as $1° = (\pi/180)$ rad, and the subdivisions of minute—$1' = (1/60)°$, and second—$1'' = (1/60)'$. The centesimal system is used in some European countries and has the *grade* as its basic unit. This is defined as $1^g = (\pi/200)$ rad. Subdivisions are the centesimal minute—$1^c = (10^{-2})^g$, and the centesimal second—$1^{cc} = (10^{-2})^c$.

It is not normally necessary to convert from one system to another. If theodolite readings in one of the practical systems are obtained then tables of trigonometrical functions with the argument expressed in

Relationship between the three systems of units of plane angular measurement

|  | Radian | Sexagesimal | Centesimal |
|---|---|---|---|
| Radian |  | 1 rad $\simeq$ 57·295 8°<br>„ $\simeq$ 3 437·75'<br>„ $\simeq$ 206 265" | 1 rad $\simeq$ 63·662 0$^g$<br>„ $\simeq$ 6 366·20$^c$<br>„ $\simeq$ 636 620$^{cc}$ |
| Sexagesimal | 1° $\simeq$ 1·745 33 $\times$ 10$^{-2}$ rad<br>1' $\simeq$ 2·908 88 $\times$ 10$^{-4}$ „<br>1" $\simeq$ 4·848 14 $\times$ 10$^{-6}$ „ |  | 1° $\simeq$ 1·111 11$^g$<br>1' $\simeq$ 1·851 85$^c$<br>1" $\simeq$ 3·086 42$^{cc}$ |
| Centesimal | 1$^g$ $\simeq$ 1·570 80 $\times$ 10$^{-2}$ rad<br>1$^c$ $\simeq$ 1·570 80 $\times$ 10$^{-4}$ „<br>1$^{cc}$ $\simeq$ 1·570 80 $\times$ 10$^{-6}$ „ | 1$^g$ $=$ 0·9°<br>1$^c$ $=$ 0·54'<br>1$^{cc}$ $=$ 0·324" |  |

units of that system should be used. In theoretical work, however, (see equation (1.7), for example) angles are expressed in radian measure. When numerical values in the centesimal system, for example, are inserted into the equations, these values must be converted to radian measure. This most commonly occurs when small angles are involved and are expressed in seconds. The preceding table shows that 1 rad $\simeq$ 206 265″ $\simeq$ 636 620$^{cc}$. Since for small angles sin $\alpha \simeq \alpha$ rad, it is often said that

$$\sin 1'' = 1/\rho'' \simeq 1/206\ 265$$

or
$$\sin 1^{cc} = 1/\rho^{cc} \simeq 1/633\ 620$$

so that
$$\alpha \text{ rad} = \alpha''/\rho'' = \alpha^{cc}/\rho^{cc}$$

### 1.2.3 *Some Other Units used in Survey Measurements*

The SI unit of *temperature* is the *Kelvin degree* (°K) and this is the unit of the thermodynamic scale of temperature. The fundamental fixed point of this scale is the triple point of water which is said to have a temperature of 273·16 °K. The temperature of the ice point is then 273·15 °K. The *Celsius degree* (°C) is a unit of temperature equal in magnitude to 1 °K. The Celsius scale of temperature is based on 0 °C = 273·15 °K. A temperature interval of $t$ °K ($= t$ °C) is often written as $t$ degC. The Celsius scale is identical to the *centigrade* scale, but the latter term was rejected by the CGPM partly to avoid confusion with the centigrade (1°) unit of plane angular measure (section 1.2.2).

The SI unit of *force* is the *newton* (N) which is the force required to give a mass of 1 kg an acceleration of 1 m/s². Thus 1 N = 1 kg m/s². The *kilogramme–force* (kgf) is the force required to give a mass of 1 kg an acceleration of 9·806 65 m/s², this acceleration being the acceleration produced by a 'standard gravity'. Thus, 1 kgf = 9·806 65 N.

The SI unit of *pressure* is the *pascal* (Pa). This is the pressure produced by the uniform application of a force of 1 N over an area of 1 square metre. Thus, 1 Pa = 1 N/m². The *standard atmosphere* (which supports a column of mercury 760 mm high) is 101·325 kN/m² or 101 325 Pa. The *bar* is defined as $10^5$ N/m² so that 1 mbar = 100 N/m² and 1 *mm Hg* $\simeq$ 1·333 22 mbar. It follows that 760 mm Hg = 1 013·25 mbar.

The SI unit of *area* is the square metre. The *hectare* (derived from hecto-are, where 1 are = 100 m²) is $10^4$ m², or 1 ha = $10^4$ m².

## 1.3  The Mathematical Model and Systematic Errors

The physical reality of the components of a survey is complex. Triangulation and traverse marks are often bolts in concrete or nails in wooden pegs; optical and electromagnetic measurements are made through an atmosphere which may be dust-laden or turbulent because of the effects of heat energy; on a construction site, vibrations from earth-moving plant or drilling equipment will affect the components of measuring instruments; tripod legs often have to be embedded in unstable, recently excavated material. All of these factors are components of the survey and affect the measurements to a greater or less extent.

Such a complex situation has to be simplified if the measurement is to give rise to values of the quantities measured. A mathematical model is constructed to represent the physical reality and values are obtained using this model. For example, in a particular model nails in pegs could become points, lines of minimum optical path become straight lines joining the points, all points remain fixed relative to each other, graduations on measuring apparatus represent the standard units, and so on.

As an illustration of a particular mathematical model, consider the measurement of a horizontal distance ($H$) between two pegs using a steel tape graduated throughout in 10 mm intervals. This could be carried out by aligning the tape between the two pegs and by reading it at the points where it coincides with the centres of the pegs. If these readings are $R_1$ and $R_2$, one mathematical model of the reality is

$$H = (R_1 \sim R_2) \tag{1.2}$$

which leads to a value for $H$, the horizontal distance. If the pegs are at different heights, the value obtained using this particular model is in error. A better model would then be

$$H = (R_1 \sim R_2) - \{\Delta h^2/2(R_1 \sim R_2)\} \tag{1.3}$$

(see section 6.1.1) where $\Delta h$ is the height difference between the pegs. One can identify further possible sources of error arising from the use of model (1.3). For example, the graduations on the tape may not represent the standard of length because of thermal expansion of the steel, or because of mechanical expansion as a result of applied tension. The mathematical model can be refined by the inclusion of further terms to take account of these sources of error if the law governing the error is known. These terms are often referred to as 'corrections'. Errors arising from an incorrect choice of mathematical model are called *systematic errors*.

The mathematical model for any particular measurement will depend on the purpose of the measurement. If the horizontal distance is required for the calculation of a cross-sectional area and subsequently of a volume of excavation, then $H = (R_1 \sim R_2)$ is probably adequate. If the distance is part of a traverse to provide control for large-scale mapping, further terms will probably be required. A trial computation of the likely magnitudes of such 'correction' terms will assist in the correct choice of a mathematical model for a particular task. An understanding of the propagation of systematic errors (section 1.3.2) is also necessary. Chapter 6 deals with some fundamental mathematical models used in basic surveying procedures.

### 1.3.1  *Reduction of Systematic Errors*
If the law governing a systematic error is known, a term in the mathematical model can reduce the error to an insignificant size. This is not always convenient. For example, the error ($e$) in the horizontal circle reading of a theodolite having a horizontal collimation error ($i$) is given by $e = i \sec h$, where ($h$) is the angle of elevation to the target. The mathematical model could include terms to allow for this source of error, but the theodolite would have to be tested frequently to determine ($i$). It is more convenient to arrange the procedure (section 1.4) so that the effect of this error can almost be eliminated: if the theodolite is reversed (i.e., a pointing made on face-left and one on face-right) the mean of the two readings will be free from this collimation error (if ($i$) remains numerically constant during the procedure), since the sign of ($i$) changes during reversal. This *principle of reversal* is often used to reduce systematic errors.

A measurement procedure can be adopted to reduce the effects of a systematic error whose law is not known. For example, in trigonometrical levelling the effect of atmospheric refraction on the observed vertical angle depends upon the meteorological conditions around the line of sight. These conditions can never be determined in normal work, so the model is inadequate (see section 6.5.1). However, by making simultaneous observations from each end of the line and averaging the results, the systematic error can be greatly reduced.

### 1.3.2  *Propagation of Systematic Errors*
Even with a well-designed mathematical model, there will be some systematic error present in the value of a measured quantity. The

effect of such an error on the calculated value of a function of the measured quantity has to be considered. Suppose $X$ is a function of $n$ independent variables $x_1, x_2,..., x_n$;

$$X = f(x_1, x_2,..., x_n) \tag{1.4}$$

If $x_1, x_2,..., x_n$ are measured and the values obtained have small systematic errors $\delta x_1, \delta x_2,..., \delta x_n$ respectively, the value of $X$ calculated from (1.4) will have an error $\delta X$ given by

$$\delta X \simeq \left(\frac{\partial f}{\partial x_1}\right) \delta x_1 + \left(\frac{\partial f}{\partial x_2}\right) \delta x_2 + \cdots + \left(\frac{\partial f}{\partial x_n}\right) \delta x_n \tag{1.5}$$

As an illustration consider the difference in northings ($n$) of a line AB of length $l = 100$ m and bearing $\phi = 30°$ (Fig 1.1). AB' − AB = $\delta l$.

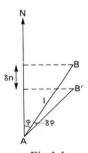

Fig 1.1

$$n = l \cos \phi \tag{1.6}$$

Suppose there is a systematic error of $+1'$ in the measured bearing ($\delta\phi = +1'$) and a systematic error of $-10$ mm in the measured length ($\delta l = -10$ mm). The presence of these errors will cause an error in the calculated northing component given by

$$\delta n \simeq \frac{\partial}{\partial\phi} (l \cos \phi).\delta\phi + \frac{\partial}{\partial l}(l \cos \phi).\delta l \tag{1.7}$$

Therefore $\qquad \delta n \simeq -l \sin \phi.\delta\phi + \cos \phi.\delta l$

Substitution of numerical values gives

$$\delta n \simeq -10^5.\sin 30°.(60)(1/\rho'') + \cos 30°(-10) \text{ mm}$$

noting that $\delta\phi$ is converted to radians by the terms $(60)(1/\rho'')$—see section 1.2.2—and that the $\simeq$ sign is used since quantities involving the product $\delta\phi \cdot \delta l$ are neglected.

Thus, $\qquad \delta n \simeq -(14\cdot5 + 8\cdot7) \simeq -23$ mm

Often it is possible to estimate the likely maximum magnitude of a residual systematic error in a measurement. The effect of such an error on derived values can then be estimated using the above method. It can then be decided whether or not it is necessary to include extra terms in the mathematical model or whether the procedure should be modified to reduce the effects of the error.

## 1.4   The Measurement Procedure and Mistakes

Having formed a mathematical model, a procedure must be prescribed for obtaining readings or observations. For example, in the measurement of the horizontal distance described in section 1.3 the procedure to be followed in the field, if $H = (R_1 \sim R_2)$ is the suitable mathematical model, will be relatively simple. It will consist of aligning the tape between the pegs, reading the tape to obtain $R_1$ and $R_2$ and recording these observations.

If a more complicated mathematical model is necessary and the effects of temperature, tension and height difference have to be allowed for, the procedure will involve the measurement and recording of temperature, the application of tension with a spring balance, for example, and levelling between the pegs, as well as tape reading and recording.

Failure to follow the prescribed procedure will give rise to an error and this type of error may be called a *mistake* or *blunder* on the part of the observer. For example, a tape reading of 38·23 m may be recorded as 38·32 m in the field-book, the thermometer may be misread, or the measurement may be made between the wrong pegs altogether. Such mistakes can be very serious if their presence is not detected at an early stage. It is desirable to arrange the procedure to detect the presence of a large mistake. An incorrect booking of a tape reading may be detected by taking two readings of $R_1$ and $R_2$ using different sections of the tape and checking that the values of $(R_1 \sim R_2)$ obtained are close together. This constitutes an *independent check* on the tape readings. An incorrect thermometer reading could be detected by the independent check of a second observer or a second thermometer, or both. If the distance being measured is part of a traverse, the traverse should be closed as an independent check. Mistakes can, of course, be avoided by care on the part of the observer.

A procedure can often be designed both to provide an independent check and also to reduce systematic error. For example, if an angle is measured once on face-left and once on face-right, the two values give a check on mistaken reading and booking (especially if the circle is rotated at change of face) and the mean of the two will be almost free of horizontal collimation error (see section 1.3.1).

Any survey procedure must include *independent* checks of this kind; the expense incurred by a failure to recognise the way in which mistakes could be detected early on in a survey is often very great.

## 1.5  Random Errors

If a quantity is measured several times under exactly the same conditions, using exactly the same procedure and the same mathematical model, the values obtained will be identical. (But note that this value need not be—and probably will not be—the 'true' value.) Such a result is impossible to achieve in practice, since a human observer cannot repeat exactly a prescribed procedure, nor can the conditions of measurement be controlled exactly. For example, if repeated measurements of the distance between two pegs are made using a steel tape, the observer's judgement of the tape readings will vary slightly from one measurement to another and so will his application of the prescribed tension. Moreover, one of the conditions of measurement, the temperature of the tape, will vary because of energy transfers between the tape and its surroundings, and it is not possible to register or control all these variations under field conditions.

Therefore, in practice, if repeated measurements of the same quantity are made under (as far as possible) the same conditions using (as far as possible) the same procedure and the same mathematical model, the values obtained will show a variation. If the variations in the procedure and in the conditions of measurement are random, then the variations in the measured values will also be random, and the values are said to have *random errors*. The difference between any two such measured values is a *discrepancy*. Analysis of values subject to random errors is based on statistical principles. Chapter 2 deals with some of these principles and Chapters 3 and 4 deal with the application of these principles to analysis and adjustment of survey measurements.

# CHAPTER TWO

# Properties of a Random Variable

## 2.1 Definition of a Random Variable (or Variate)

A random variable is one whose value depends upon chance. In statistics, such a variable is often called *stochastic* ($\sigma\tau o\chi\acute{a}\zeta\epsilon\sigma\theta\alpha\iota =$ to guess). A set of measured values $x_1, x_2, ..., x_n$ obtained 'under the same conditions' (section 1.5) constitutes a *sample* of the random variable, $x$, which is being measured. If the measurements were to be repeated an infinite number of times, the resulting infinite number of values $x_1, x_2, ...$ would constitute the *population* of the random variable $x$.

In practical surveying, the size of a sample is generally small; seldom are more than ten repeated measurements made. From this sample we have to deduce characteristics of the population. From these population characteristics (or parameters) we can make deductions about the relative reliability of different measurements and about the most likely value of the random variable $x$.

## 2.2 The Cumulative Distribution Function (CDF)

Suppose that $x$ is a random variable and that $t$ is a parameter which can vary from $-\infty$ to $\infty$. Denote the probability of the value $x \leqslant t$ occurring in a measurement as $P(x \leqslant t)$. The cumulative distribution function (CDF) of $x$ is defined as

$$F(t) = P(x \leqslant t) \tag{2.1}$$

Then, as $t \to -\infty$, $F(t) \to 0$ and as $t \to \infty$, $F(t) \to 1$.

For $a < b$, $\qquad F(b) - F(a) = P(a < x \leqslant b) \tag{2.2}$

If $x$ can take only discrete values $t_1, t_2, t_3,\ldots$, then the CDF is a step function. A step CDF is illustrated in Fig. 2.1(a) where the 'vertical step' at $t = t_3$ is $F(t_3) - F(t_2)$, which from (2.2) is $P(t_2 < x \leqslant t_3)$ or $P(x = t_3)$ since $x$ can take only discrete values $t_1, t_2, t_3,\ldots$.

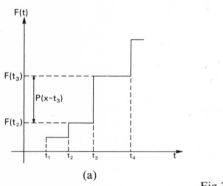

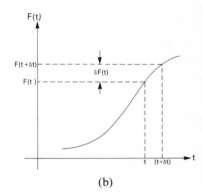

(a) (b)

Fig 2.1

In surveying, the distances and angles which are measured are taken as being continuous variables and the CDF is assumed to be continuously differentiable, with the parameter $t$ continuous from the right. Fig 2.1(b) illustrates a continuous CDF, where

$$\delta F(t) \simeq \frac{\mathrm{d}F(t)}{\mathrm{d}t} \cdot \delta t = f(t)\, \delta t \tag{2.3}$$

Then, $\qquad P(t < x \leqslant t + \delta t) = \delta F(t) \simeq f(t)\, \delta t$

## 2.3 The Probability Density (PD)

The function $f(t)$ defined in (2.3) as

$$f(t) = \frac{\mathrm{d}F(t)}{\mathrm{d}t}$$

is called the *probability density* (PD) of the random variable $x$. A PD curve is illustrated in Fig 2.2.

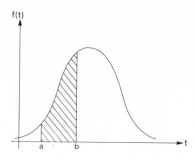

Fig 2.2

From (2.2),     $F(b) - F(a) = P(a < x \leqslant b)$

Therefore     $$P(a < x \leqslant b) = \int_a^b f(t)\, dt \qquad (2.4)$$

which is equal to the area of the shaded portion in Fig 2.2. Also, since the probability that $x$ lies between $-\infty$ and $\infty$ is 1 (i.e. it is certain):

$$\int_{-\infty}^{\infty} f(t)\, dt = 1 \qquad (2.5)$$

The CDF and PD are sometimes expressed as functions of the variable $x$ itself rather than in terms of a parameter $t$, i.e. as $F(x)$ and $f(x)$ respectively. The remainder of this chapter deals with properties of a random variable, assuming only that the CDF and PD exist. It is not necessary at this stage to define the form of either function, but only to assume that the PD exists as a result of the CDF being continuously differentiable.

## 2.4   The Expectation of a Random Variable

If a random variable $x$ has a continuously differentiable CDF, $F(x)$, and a PD, $f(x)$, the *expectation* of $x$ (written as $E\{x\}$) is the *mean value* taken over the population, i.e.

$$E\{x\} = \int_{-\infty}^{\infty} x f(x)\, dx \bigg/ \int_{-\infty}^{\infty} f(x)\, dx = \int_{-\infty}^{\infty} x f(x)\, dx \qquad (2.6)$$

If $x$ has a step CDF $F(x)$, then its expectation is

$$E\{x\} = \sum_{i=-\infty}^{\infty} x_i p_i \bigg/ \sum_{i=-\infty}^{\infty} p_i = \sum_{i=-\infty}^{\infty} x_i p_i \qquad (2.7)$$

where $p_i$ is the probability of a measurement $x_i$. Equations (2.6) and (2.7) are analogous to the equations defining the centroid of a continuous rigid body (2.6) and of a system of particles (2.7).

It can be shown (see, for example, [13]) that for $n$ random variables $x_1, x_2, ..., x_n$

$$E\{x_1 + x_2 + \cdots + x_n\} = E\{x_1\} + E\{x_2\} + \cdots + E\{x_n\} \qquad (2.8)$$

and that for a constant $c$,

$$E\{cx\} = c.E\{x\} \qquad (2.9)$$

If $y$ is defined uniquely by $y = \phi(x)$ where $x$ is a random variable with PD $f(x)$, then $y$ is also a random variable and its expectation is given by

$$E\{y\} = E\{\phi(x)\} = \int_{-\infty}^{\infty} \phi(x) f(x) \, dx \qquad (2.10)$$

If $x$ has a step CDF and $y = \phi(x)$ as above, then

$$E\{y\} = E\{\phi(x)\} = \sum_{i=-\infty}^{\infty} \phi(x_i) p_i \qquad (2.11)$$

where $p_i$ is the probability that a value $x = x_i$ is obtained.

## 2.5 Variance and Standard Deviation

If a random variable $x$ has expectation $\xi$, its *variance* (written as $V\{x\}$, or $\sigma_x^2$) is defined as

$$V\{x\} = E\{(x - \xi)^2\} = \sigma_x^2 \qquad (2.12)$$

Thus, if $x$ has a PD $f(x)$, it follows from (2.10) that

$$V\{x\} = \sigma_x^2 = \int_{-\infty}^{\infty} (x - \xi)^2 f(x) \, dx \qquad (2.13)$$

and if $x$ has a step CDF $F(x)$, it follows from (2.11) that

$$V\{x\} = \sigma_x^2 = \sum_{i=-\infty}^{\infty} (x_i - \xi)^2 p_i \qquad (2.14)$$

where $p_i$ is the probability that $x = x_i$.

The *standard deviation* (SD) of a random variable $x$ is defined as the positive square root of the variance and is written as $\sigma_x$.

When the random variable $x$ is a quantity which has been measured several times 'under the same conditions', the variance gives an indication of the *precision* of the measurements: the precision is inversely proportional to the variance. The fundamental difference between precision and accuracy is discussed in section 3.3.

## 2.6  Multivariate Distributions

If $x$ and $y$ are two random variables (for example the measured bearing and distance from one point to another) and $t$ and $u$ are parameters which can vary from $-\infty$ to $\infty$, the (bivariate) CDF of $x$ and $y$ is defined as

$$F(t, u) = P(x \leqslant t, y \leqslant u) \qquad (2.15)$$

where $F(t, u)$ is either a step function or is continuously differentiable. In the latter case (illustrated in Fig 2.3) the PD of $x$ and $y$ is defined as

$$f(t, u) = \frac{\partial^2 F(t, u)}{\partial t\, \partial u} \qquad (2.16)$$

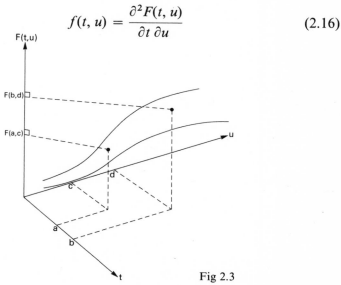

Fig 2.3

Then, $\quad P(t < x \leqslant t + \delta t, u < y \leqslant u + \delta u) \simeq f(t, u)\, \delta t\, \delta u \quad$ (2.17)

and

$$P(a < x \leqslant b, c < y \leqslant d) = F(b, d) - F(a, c) = \int_a^b \int_c^d f(t, u)\, dt\, du$$
(2.18)

Also, $\qquad\qquad \int_{-\infty}^{\infty} \int_{-\infty}^{\infty} f(t, u)\, dt\, du = 1 \qquad\qquad$ (2.19)

The (univariate) PD curve of Fig 2.2 becomes a PD surface in the case of a bivariate distribution (Fig 2.4) and an interpretation of (2.18) is that the volume under the surface bounded by $t = a, b$ and $u = c, d$ is the probability that $a < x \leqslant b$ and $c < y \leqslant d$.

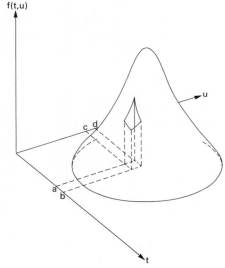

f(t,u)

Fig 2.4

The functions $f(t, u)$ and $F(t, u)$ are often written in terms of the (continuous) random variables $x$ and $y$ as $f(x, y)$ and $F(x, y)$ respectively.

The two random variables $x$ and $y$ are *statistically independent* if

$$F(x, y) = G(x)H(y) \qquad\qquad (2.20)$$

where $F(x, y)$ is the CDF of $x$ and $y$, $G(x)$ is the CDF of $x$, and $H(y)$ is the CDF of $y$.

In such a case,

$$f(x, y) = \frac{\partial^2 F(x, y)}{\partial x\, \partial y} = \frac{\partial^2 G(x)H(y)}{\partial x\, \partial y} = \frac{dG(x)}{dx} \cdot \frac{dH(y)}{dy}$$

or
$$f(x, y) = g(x)h(y) \qquad (2.21)$$

where $g(x)$ is the PD of $x$ and $h(y)$ is the PD of $y$. As a result,

$$P(x_1 < x \leqslant x_2, y_1 < y \leqslant y_2) = \int_{x_1}^{x_2} \int_{y_1}^{y_2} f(x, y)\, dx\, dy$$

$$= \int_{x_1}^{x_2} g(x)\, dx \int_{y_1}^{y_2} h(y)\, dy$$

$$= P(x_1 < x \leqslant x_2) . P(y_1 < y \leqslant y_2)$$

$$(2.22)$$

Further, it can be shown that if $x$ and $y$ are independent random variables,

$$E\{xy\} = E\{x\} . E\{y\} \qquad (2.23)$$

For $n$ random variables $x_1, x_2, \ldots, x_n$ the PD is

$$f(x_1, x_2, \ldots, x_n) = \frac{\partial^n F(x_1, x_2, \ldots, x_n)}{\partial x_1, \partial x_2, \ldots, \partial x_n}$$

## 2.7  Covariance

The variance of the sum $(z)$ of two random variables $x$ and $y$ with expectations $\xi$ and $\eta$ respectively is defined as

$$V\{x + y\} = V\{z\} = E\{(x - \xi + y - \eta)^2\}$$

$$= E\{(x - \xi)^2 + (y - \eta)^2 + 2(x - \xi)(y - \eta)\}$$

Therefore, from (2.8) and (2.9)

$$V\{z\} = V\{x\} + V\{y\} + 2E\{(x - \xi)(y - \eta)\} \qquad (2.24)$$

The expectation of $(x - \xi)(y - \eta)$ is the *covariance* of $x$ and $y$. If the variance of $z$ is written as $\sigma_z^2$, the variances of $x$ and $y$ are written as $\sigma_x^2$ and $\sigma_y^2$ respectively and the covariance of $x$ and $y$ is written as $\sigma_{xy}$, equation (2.24) becomes

$$\sigma_z^2 = \sigma_x^2 + \sigma_y^2 + 2\sigma_{xy} \qquad (2.25)$$

An alternative nomenclature for equations (2.24) and (2.25) is

$$\text{var}\,(x \pm y) = \text{var}\,(x) + \text{var}\,(y) \pm 2\,\text{cov}\,(xy) \qquad (2.26)$$

If $x$ and $y$ are independent random variables, their covariance can be shown to be zero:

$$\begin{aligned}
\sigma_{xy} &= E\{(x - \xi)(y - \eta)\} = E\{xy - x\eta - \xi y + \xi\eta\} \\
&= E\{xy\} - \eta E\{x\} - \xi E\{y\} + \xi\eta \\
&= E\{x\}E\{y\} - \eta\xi - \xi\eta + \xi\eta \\
&= 0
\end{aligned}$$

Equation (2.25) can be extended to express the variance $(\sigma_x^2)$ of the sum of $n$ *independent* random variables $x_1, x_2, ..., x_n$ with respective variances $\sigma_{x_1}^2, \sigma_{x_2}^2, ..., \sigma_{x_n}^2$ :

$$\sigma_x^2 = \sigma_{x_1}^2 + \sigma_{x_2}^2 + \cdots + \sigma_{x_n}^2 \qquad (2.27)$$

## 2.8   Correlation

The covariance of two random variables $x$ and $y$ has been defined (section 2.7) as

$$\sigma_{xy} = E\{(x - \xi)(y - \eta)\} \qquad (2.28)$$

In order to examine the meaning of this in a surveying context, consider the measurements of the lengths $x$ and $y$ of two adjacent legs in a traverse using a steel tape. We shall assume that the mathematical model for $x$ is $(R_2 - R_1)$ and that for $y$ is $(R_2' - R_1')$ where the $R$s denote tape readings.

In the first instance, suppose that repeated measurements of $x$ and $y$ are made in rapid succession and, as far as possible, under the same conditions. These repeated measurements give rise to a sample of each of the random variables $x$ and $y$, and, since the measurements are made rapidly, there is no overall rise or fall in temperature.

In the second instance, suppose that one measurement of $x$ and one of $y$ are made in rapid succession and an hour later another pair of measurements is made, and so on for many days. Once again, these measurements will give rise to a sample of each of the random variables $x$ and $y$. This time, however, any pair of measurements will be made at a temperature which will be significantly different from that of some other pairs of measurements: at high temperatures, both

$x$ and $y$ will tend to have lower values, and at low temperatures, $x$ and $y$ will tend to have higher values. In such a case, $x$ and $y$ are said to be (positively) correlated. In the first instance, however, there is no reason to suppose that the measures will fall into such an overall pattern, and $x$ and $y$ are uncorrelated.

The *coefficient of correlation* between two random variables is defined as

$$\rho_{xy} = \sigma_{xy}/\sigma_x\sigma_y \tag{2.29}$$

where $-1 \leqslant \rho_{xy} \leqslant 1$. When $\rho_{xy} = \pm 1$, the variables are completely correlated (positively or negatively); when $\rho_{xy} = 0$, the variables are uncorrelated, or independent.

In surveying, there are often no grounds for supposing the measurements to be correlated. However, some correlation exists between quantities which are computed from the same samples of random variables, since these quantities are not independent.

## 2.9  The Expectation, Variance and Covariance of Functions of Random Variables

In surveying, the measurements are usually used to compute values of functions of the measurements. It is these derived quantities (such as coordinates) which are of interest, and their expectations and variances are required. It is therefore necessary to see how the expectations and variances of the derived quantities are related to the corresponding characteristics of the measured quantities.

### 2.9.1  *Linear Functions of One Variable*

If $x$ is a random variable with expectation $\xi$ and variance $\sigma_x^2$ and if $y$ is defined by $y = ax + b$, where $a$ and $b$ are constants, the expectation ($\eta$) of $y$ is given by

$$\eta = E\{y\} = E\{ax + b\}$$

From (2.8) and (2.9):

$$\eta = a.E\{x\} + b = a\xi + b \tag{2.30}$$

The variance ($\sigma_y^2$) of $y$ is given by

$$\sigma_y^2 = V\{y\} = E\{(y - \eta)^2\} = E\{(ax + b - a\xi - b)^2\}$$

Therefore $\qquad \sigma_y^2 = E\{(ax - a\xi)^2\} = a^2 E\{(x - \xi)^2\}$

Therefore
$$\sigma_y^2 = a^2 \sigma_x^2 \qquad (2.31)$$

If $y_1 = a_1 x + b$ and $y_2 = a_2 x + b$ where $y_1$ and $y_2$ have variances $\sigma_{y_1}^2$, $\sigma_{y_2}^2$ and expectations $\eta_1$ and $\eta_2$ respectively, the covariance of $y_1$ and $y_2$ is given by (2.28) as

$$
\begin{aligned}
\sigma_{y_1 y_2} &= E\{(y_1 - \eta_1)(y_2 - \eta_2)\} \\
&= E\{(a_1 x + b_1 - a_1 \xi - b_1)(a_2 x + b_2 - a_2 \xi - b_2)\} \\
&= E\{a_1(x - \xi) a_2(x - \xi)\} \\
&= a_1 a_2 E\{(x - \xi)^2\}
\end{aligned}
$$

Therefore
$$\sigma_{y_1 y_2} = a_1 a_2 \sigma_x^2 \qquad (2.32)$$

### 2.9.2 Linear Functions of more than One Variable

If $z = x + y$, where $x$ and $y$ are random variables with expectations $\xi$ and $\eta$ respectively, the expectation ($\zeta$) of $z$ is, from (2.8), given by

$$\zeta = \xi + \eta \qquad (2.33)$$

If the variances of $x$, $y$ and $z$ are $\sigma_x^2$, $\sigma_y^2$ and $\sigma_z^2$ respectively and if the covariance of $x$ and $y$ is $\sigma_{xy}$, then, from (2.25)

$$\sigma_z^2 = \sigma_x^2 + \sigma_y^2 + 2\sigma_{xy} \qquad (2.34)$$

Equations (2.33) and (2.34) can be generalised. If

$$z = a_1 x_1 + a_2 x_2 + \cdots + a_n x_n$$

where $x_i$ ($i = 1, 2, \ldots, n$) is a random variable with expectation $\xi_i$ and variance $\sigma_{x_i}^2$ and $a_i$ is a constant, then

$$E\{z\} = \zeta = a_1 \xi_1 + a_2 \xi_2 + \cdots + a_n \xi_n \qquad (2.35)$$

and

$$
\begin{aligned}
V\{z\} = \sigma_z^2 = a_1^2 \sigma_{x_1}^2 + \cdots + a_n^2 \sigma_{x_n}^2 + 2a_1 a_2 \sigma_{x_1 x_2} + 2a_1 a_3 \sigma_{x_1 x_3} \\
+ \cdots + 2a_{n-1} a_n \sigma_{x_{n-1} x_n} \qquad (2.36)
\end{aligned}
$$

If $z_1 = a_1 x + b_1 y$ and $z_2 = a_2 x + b_2 y$ where $z_1$ and $z_2$ have variances $\sigma_{z_1}^2$, $\sigma_{z_2}^2$ and expectations $\zeta_1$ and $\zeta_2$ respectively, the covariance of $z_1$ and $z_2$ is given by (2.28) as

$$
\begin{aligned}
\sigma_{z_1 z_2} &= E\{(z_1 - \zeta_1)(z_2 - \zeta_2)\} \\
&= E\{(a_1 x + b_1 y - a_1 \xi - b_1 \eta)(a_2 x + b_2 y - a_2 \xi - b_2 \eta)\} \\
&= E\{[a_1(x - \xi) + b_1(y - \eta)][a_2(x - \xi) + b_2(y - \eta)]\} \\
&= a_1 a_2 E\{(x - \xi)^2\} + b_1 b_2 E\{(y - \eta)^2\} \\
&\quad + a_1 b_2 E\{(x - \xi)(y - \eta)\} + a_2 b_1 E\{(y - \eta)(x - \xi)\}
\end{aligned}
$$

Therefore

$$\sigma_{z_1 z_2} = a_1 a_2 \sigma_x^2 + b_1 b_2 \sigma_y^2 + (a_1 b_2 + a_2 b_1)\sigma_{xy} \qquad (2.37)$$

### 2.9.3 Non-linear Functions of One Variable

If $y = \phi(x)$ is a non-linear function of the random variable $x$, and $x$ has expectation $\xi$ and variance $\sigma_x^2$, Taylor's theorem gives

$$y = \phi(\xi) + (x - \xi)\phi'(\xi) + \tfrac{1}{2}(x - \xi)^2 \phi''(\xi) + \cdots$$

If $\phi(x)$ is *approximately* linear over the range of values of $x$, then,

$$y \simeq \phi(\xi) + (x - \xi)\phi'(\xi)$$

Therefore $E\{y\} = \eta \simeq \phi(\xi) + \phi'(\xi)E\{x - \xi\}$

Therefore
$$\eta \simeq \phi(\xi) \qquad (2.38)$$

Also, using (2.31)

$$V\{y\} = \sigma_y^2 \simeq [\phi'(\xi)]^2 \sigma_x^2$$

Therefore
$$\sigma_y^2 \simeq \left(\frac{d\phi}{dx}\right)^2_{x=\xi} \cdot \sigma_x^2 \qquad (2.39)$$

If $y_1 = \phi_1(x)$ and $y_2 = \phi_2(x)$, the covariance of $y_1$ and $y_2$ is

$$\sigma_{y_1 y_2} = E\{(y_1 - \eta_1)(y_2 - \eta_2)\}$$
$$\simeq E\{[\phi_1(\xi) + (x - \xi)\phi'_1(\xi) - \phi_1(\xi)]$$
$$\times [\phi_2(\xi) + (x - \xi)\phi'_2(\xi) - \phi_2(\xi)]\}$$
$$\simeq E\{(x - \xi)^2 \phi'_1(\xi)\phi'_2(\xi)\}$$

Therefore
$$\sigma_{y_1 y_2} \simeq \phi'_1(\xi)\phi'_2(\xi)\sigma_x^2 \qquad (2.40)$$

### 2.9.4 Non-linear Functions of more than One Variable

If $z = \phi(x, y)$ where $x$ and $y$ are random variables with expectations $\xi$ and $\eta$, Taylor's theorem gives

$$z \simeq \phi(\xi, \eta) + (x - \xi)\left(\frac{\partial \phi}{\partial x}\right)_{x=\xi} + (y - \eta)\left(\frac{\partial \phi}{\partial y}\right)_{y=\eta} \qquad (2.41)$$

if $z$ is approximately linear over the practical ranges of $x$ and $y$. Therefore, using (2.35)

$$E\{z\} = \zeta \simeq \phi(\xi, \eta) \qquad (2.42)$$

If $x$ and $y$ have variances $\sigma_x^2$ and $\sigma_y^2$ respectively and covariance $\sigma_{xy}$, then, from (2.41) and (2.36),

$$V\{z\} = \sigma_z^2 \simeq \left(\frac{\partial \phi}{\partial x}\right)_{x=\xi} \cdot \sigma_x^2 + \left(\frac{\partial \phi}{\partial y}\right)_{y=\eta} \cdot \sigma_y^2 + 2\left(\frac{\partial \phi}{\partial x}\right)_{x=\xi}\left(\frac{\partial \phi}{\partial y}\right)_{y=\eta} \cdot \sigma_{xy}$$

$$(2.43)$$

Equations (2.42) and (2.43) can be generalised, for if

$$z = \phi(x_1, x_2,..., x_n)$$

where $x_i$ ($i = 1, 2,..., n$) is a random variable with expectation $\xi_i$ and variance $\sigma_{x_i}^2$

$$E\{z\} = \xi \simeq \phi(\xi_1, \xi_2,..., \xi_n) \tag{2.44}$$

and

$$V\{z\} = \sigma_z^2 \simeq \left(\frac{\partial \phi}{\partial x_1}\right)^2_{x_1=\xi_1} \cdot \sigma_{x_1}^2 + \left(\frac{\partial \phi}{\partial x_2}\right)^2_{x_2=\xi_2} \cdot \sigma_{x_2}^2 + \cdots$$

$$+ \left(\frac{\partial \phi}{\partial x_n}\right)^2_{x_n=\xi_n} \cdot \sigma_{x_n}^2 + 2\left(\frac{\partial \phi}{\partial x_1}\right)_{x_1=\xi_1}\left(\frac{\partial \phi}{\partial x_2}\right)_{x_2=\xi_2} \cdot \sigma_{x_1 x_2} + \cdots$$

$$+ 2\left(\frac{\partial \phi}{\partial x_{n-1}}\right)_{x_{n-1}=\xi_{n-1}}\left(\frac{\partial \phi}{\partial x_n}\right)_{x_n=\xi_n} \cdot \sigma_{x_{n-1} x_n} \tag{2.45}$$

where $\sigma_{x_i x_j}$ is the covariance of $x_i$ and $x_j$ ($j = 1, 2,..., n, j \neq i$).

The relationships expressed in (2.44) and (2.45) are of great importance in analysis of surveying measurements and it can be seen that results in sections 2.9.1 to 2.9.3 are special cases of these general relationships.

Also, if $z_1 = \phi_1(x_1, x_2,..., x_n)$ and $z_2 = \phi_2(x_1, x_2,..., x_n)$, the covariance of $z_1$ and $z_2$ is

$$\sigma_{z_1 z_2} \simeq \left(\frac{\partial \phi_1}{\partial x_1} \cdot \frac{\partial \phi_2}{\partial x_1}\right)\sigma_{x_1}^2 + \cdots + \left(\frac{\partial \phi_1}{\partial x_n} \cdot \frac{\partial \phi_2}{\partial x_n}\right)\sigma_{x_n}^2$$

$$+ \left(\frac{\partial \phi_1}{\partial x_1} \cdot \frac{\partial \phi_2}{\partial x_2} + \frac{\partial \phi_1}{\partial x_2} \cdot \frac{\partial \phi_2}{\partial x_1}\right)\sigma_{x_1 x_2} + \cdots$$

$$+ \left(\frac{\partial \phi_1}{\partial x_{n-1}} \cdot \frac{\partial \phi_2}{\partial x_n} + \frac{\partial \phi_1}{\partial x_n} \cdot \frac{\partial \phi_2}{\partial x_{n-1}}\right)\sigma_{x_{n-1} x_n} \tag{2.46}$$

It can be seen that (2.40), (2.37) and (2.32) are special cases of (2.46). The use of matrix notation enables expressions such as (2.46) to be written more compactly. If $z = \mathbf{a}^\top\mathbf{x}$ where $\mathbf{a}^\top = (a_1, a_2,..., a_n)$ and $\mathbf{x}^\top = (x_1, x_2,..., x_n)$, then (2.36), for example, can be written as

$$\sigma_z^2 = \mathbf{a}^\top\boldsymbol{\sigma}_{xx}\mathbf{a} \tag{2.36a}$$

where

$$\boldsymbol{\sigma}_{xx} = \begin{pmatrix} \sigma_{x_1}^2 & \sigma_{x_1 x_2} & \cdots & \sigma_{x_1 x_n} \\ \sigma_{x_2 x_1} & \sigma_{x_2}^2 & \cdots & \sigma_{x_2 x_n} \\ \vdots & \vdots & & \vdots \\ \sigma_{x_n x_1} & \sigma_{x_n x_2} & \cdots & \sigma_{x_n}^2 \end{pmatrix}$$

is the *variance–covariance matrix of* $\mathbf{x}$.

Also, if $z_1 = \mathbf{a}^\top\mathbf{x}$ and $z_2 = \mathbf{b}^\top\mathbf{x}$, then

$$\sigma_{z_1 z_2} = \mathbf{a}^\top\boldsymbol{\sigma}_{xx}\mathbf{b} \tag{2.37a}$$

and equivalent expressions can be written for (2.45) and (2.46).


## 2.10   Sample Mean and Sample Variance

In surveying measurement, one obtains only a small sample of a random variable $x$. It is possible to deduce estimates of the expectation ($\xi$) and variance ($\sigma_x^2$) of the population from the sample.

If a random sample consists of $n$ independent measured values $x_1, x_2,..., x_n$, the arithmetic mean ($\bar{x}$) of the sample is defined as

$$\bar{x} = (1/n) \sum_{i=1}^{n} x_i \tag{2.47}$$

and $\bar{x}$ is itself a random variable with expectation

$$E\{\bar{x}\} = (1/n)E\left\{ \sum_{i=1}^{n} x_i \right\}$$

and from (2.35),

$$E\{\bar{x}\} = (1/n)(\xi + \xi + \cdots + \xi) = \xi$$

Thus the expectation of the arithmetic mean of the sample is the expectation of the population. The sample mean is thus said to be an *unbiased estimator* of the population mean.

The variance of $\bar{x}$ from (2.36) is given by

$$\sigma_{\bar{x}}^2 = (\sigma_x^2/n^2) + (\sigma_x^2/n^2) + \cdots + (\sigma_x^2/n^2) = \sigma_x^2/n \qquad (2.48)$$

so the standard deviation of the mean ($\sigma_{\bar{x}}$) is given by

$$\sigma_{\bar{x}} = \sigma_x/\sqrt{n} \qquad (2.49)$$

The variance of $x$ is defined as

$$V\{x\} = \sigma_x^2 = E\{(x - \xi)^2\}$$

and an approximation to $\sigma_x^2$ from (2.14) is given by

$$s_0^2 = (1/n) \sum_{i=1}^{n} (x_i - \xi)^2 \qquad (2.50)$$

It can be seen that $E\{s_0^2\} = \sigma_x^2$. However, since, in practice, the population mean $\xi$ is unknown, we can use the sample mean $\bar{x}$ instead of $\xi$ in (2.50) and obtain

$$s^2 = (1/n) \sum_{i=1}^{n} (x_i - \bar{x})^2$$

But in this case $s^2$ is not an unbiased estimator of $\sigma_x^2$, since $E\{s^2\} \neq E\{s_0^2\} = \sigma_x^2$. Form instead

$$s_x^2 = (1/k) \sum_{i=1}^{n} (x_i - \bar{x})^2 \qquad (2.51)$$

where $k$ is to be determined so that $E\{s_x^2\} = \sigma_x^2$. Write $(x_i - \bar{x})^2$ as $[(x_i - \xi) - (\bar{x} - \xi)]$. Then

$$\sum_{i=1}^{n} (x_i - \bar{x})^2 = \sum_{i=1}^{n} [(x_i - \xi) - (\bar{x} - \xi)]^2$$

$$= \sum_{i=1}^{n} (x_i - \xi)^2 - 2 \sum_{i=1}^{n} (x_i - \xi)(\bar{x} - \xi) + n(\bar{x} - \xi)^2$$

$$= \sum_{i=1}^{n} (x_i - \xi)^2 - 2(\bar{x} - \xi)(n\bar{x} - n\xi) + n(\bar{x} - \xi)^2$$

$$= \sum_{i=1}^{n} (x_i - \xi)^2 - n(\bar{x} - \xi)^2$$

Substituting in (2.51), we have

$$s_x^2 = (1/k)\left[\sum_{i=1}^n (x_i - \xi)^2 - n(\bar{x} - \xi)^2\right]$$

Thus, if $s_x^2$ is to be an unbiased estimator of $\sigma_x^2$,

$$E\{s_x^2\} = (1/k)E\left\{\sum_{i=1}^n (x_i - \xi)^2\right\} - (n/k)E\{(\bar{x} - \xi)^2\} = \sigma_x^2$$

$$\therefore \quad (1/k)n\sigma_x^2 - (n/k)\sigma_x^2/n = \sigma_x^2$$

$$\therefore \quad k = (n - 1)$$

Thus
$$s_x^2 = \frac{1}{n - 1} \sum_{i=1}^n (x_i - \bar{x})^2$$

and
$$s_x = \sqrt{\frac{\sum\limits_{i=1}^n (x_i - \bar{x})^2}{(n - 1)}} \tag{2.52}$$

are unbiased estimators of the variance and standard deviation respectively of the random variable $x$. It can be shown, similarly, that unbiased estimators of the variance and of the standard deviation of the sample mean $\bar{x}$ are, respectively,

$$s_{\bar{x}}^2 = \frac{\sum\limits_{i=1}^n (x_i - \bar{x})^2}{n(n - 1)}$$

and
$$s_{\bar{x}} = \sqrt{\frac{\sum\limits_{i=1}^n (x_i - \bar{x})^2}{n(n - 1)}} \tag{2.53}$$

and that the unbiased estimator of the covariance of two random variables $x$ and $y$ is

$$s_{xy} = \frac{\sum\limits_{i=1}^n (x_i - \bar{x})(y_i - \bar{y})}{n - 1} \tag{2.54}$$

## 2.11   Standard Errors

The unbiased estimate of the standard deviation of a random variable $x$ (defined in (2.52)) is commonly called the *standard error of a measurement*, $x_i$. The use of this term is widespread, although the quantity referred to is not an error as defined in (1.1). Similarly, $s_{\bar{x}}$, defined in (2.53), is called the *standard error of the mean* of the measurements. These two quantities, together with $s_{xy}$ (defined in (2.54)), are in common use in the analysis of survey and other measurements. Although the expectation and the variance are the parameters which describe a population, in surveying and other measurement sciences samples of the population are obtained, and the mean and standard error are often taken to be identical with expectation and standard deviation, respectively.

The propagation of means and standard errors is analogous to that of expectations and variances. Equations (2.30) to (2.46) in section 2.9 can be used to find the means and standard errors of functions of measured quantities. Thus, in general, if

$$z = \phi(x_1, x_2, \ldots, x_n)$$

where $x_i$ ($i = 1, 2, \ldots, n$) has standard error $s_i$ and mean $\bar{x}_i$,

$$\bar{z} = \phi(\bar{x}_1, \bar{x}_2, \ldots, \bar{x}_n) \tag{2.55}$$

and the standard error of $z$ ($s_z$) is given by

$$s_z^2 = \left(\frac{\partial \phi}{\partial x_1}\right)^2 . s_1^2 + \left(\frac{\partial \phi}{\partial x_2}\right)^2 . s_2^2 + \cdots + \left(\frac{\partial \phi}{\partial x_n}\right)^2 . s_n^2$$

$$+ 2\left(\frac{\partial \phi}{\partial x_1}\right)\left(\frac{\partial \phi}{\partial x_2}\right)s_{1,2} + 2\left(\frac{\partial \phi}{\partial x_1}\right)\left(\frac{\partial \phi}{\partial x_3}\right)s_{1,3} + \cdots$$

$$+ 2\left(\frac{\partial \phi}{\partial x_{n-1}}\right)\left(\frac{\partial \phi}{\partial x_n}\right)s_{n-1,n} \tag{2.56}$$

where $s_{ij}$ is the unbiased estimate of the covariance of $x_i$ and $x_j$, and $(\partial \phi/\partial x_i)$ is evaluated for $x_i = \bar{x}_i$.

Also, if $z_1 = \phi_1(x_1, \ldots, x_n)$ and $z_2 = \phi_2(x_1, \ldots, x_n)$, the estimate of the covariance of $z_1$ and $z_2$ is

$$s_{z_1 z_2} = \left( \frac{\partial \phi_1}{\partial x_1} \cdot \frac{\partial \phi_2}{\partial x_1} \right) s_{x_1}^2 + \cdots + \left( \frac{\partial \phi_1}{\partial x_n} \cdot \frac{\partial \phi_2}{\partial x_n} \right) s_{x_n}^2$$

$$+ \left( \frac{\partial \phi_1}{\partial x_1} \cdot \frac{\partial \phi_2}{\partial x_2} + \frac{\partial \phi_1}{\partial x_2} \cdot \frac{\partial \phi_2}{\partial x_1} \right) s_{x_1 x_2}$$

$$+ \left( \frac{\partial \phi_1}{\partial x_1} \cdot \frac{\partial \phi_2}{\partial x_3} + \frac{\partial \phi_1}{\partial x_3} \cdot \frac{\partial \phi_2}{\partial x_1} \right) s_{x_1 x_3}$$

$$+ \cdots + \left( \frac{\partial \phi_1}{\partial x_{n-1}} \cdot \frac{\partial \phi_2}{\partial x_n} + \frac{\partial \phi_1}{\partial x_n} \cdot \frac{\partial \phi_2}{\partial x_{n-1}} \right) s_{x_{n-1} x_n} \qquad (2.57)$$

# CHAPTER THREE

# The Principle of Least Squares

## 3.1 Legendre's Postulate

In 1806 Legendre [19] proposed a method for obtaining the most probable value (MPV) of a quantity, given a set of $n$ equally reliable measured values $(x_1, x_2, ..., x_n)$ of the quantity. According to Legendre, the MPV $(x)$ is that which makes the sum of the squares of the residuals a minimum, a *residual* $(v_i)$ being defined as

$$v_i = (x - x_i) \tag{3.1}$$

It follows from this postulate that the arithmetic mean $(\bar{x})$ of a series of equally reliable observations is the MPV, as the following reasoning shows: the sum of the squares of the residuals is

$$\sum_{i=1}^{n} (x - x_i)^2 = \sum_{i=1}^{n} v_i^2 = [v_i^2]$$

(This notation of square brackets to denote summation over the finite range $i = 1$ to $n$ is common in least squares (LS) literature.) Thus, according to the LS principle

$$(d/dx)[v_i^2] = 0, \quad \text{i.e.} \quad (d/dx)[(x - x_i)^2] = 0$$

therefore

$$2nx - 2[x_i] = 0$$

and

$$x = [x_i]/n = \bar{x}$$

noting that $(d^2/dx^2)\,[v_i^2] = 2n > 0$ indicates a minimum.

## 3.2 The Theoretical Basis of Least Squares

Gauss [15] attempted to place the LS principle on a logical foundation. Starting from the postulate that when a number of equally reliable independent measurements of an unknown quantity are made, the MPV is the arithmetic mean, Gauss deduced that the measurements must be distributed according to the *normal frequency distribution*. For a demonstration of Gauss's proof, see [25].

Measurements which are normally distributed have a PD (section 2.3) given by

$$f(x) = (\sigma_x\sqrt{2\pi})^{-1} \exp - (x - \xi)^2/2\sigma_x^2 \tag{3.2}$$

where $\xi$ is the expectation of $x$ and $\sigma_x^2$ is the variance of $x$. The constant term $(\sigma_x\sqrt{2\pi})^{-1}$ makes

$$\int_{-\infty}^{\infty} f(x)\,\mathrm{d}x = 1$$

according to (2.5).

Gauss and Laplace (see [25] for a discussion of their work) made other attempts to justify the LS method, without resorting to the postulate of the arithmetic mean. These works give the derivation that the MPV of an unknown quantity is that which makes the variance of the measurements a minimum. More recent work has placed the LS principle within the context of Fisher's *method of maximum likelihood*, with its application based on the *central limit-theorem* of statistics (see, for example, [16]).

Whether or not the observations are normally distributed and independent, the LS principle provides a systematic and relatively simple method of assigning values to unknown quantities when the number of measurements is greater than the number of unknowns. No other method has yet been found superior in these respects.

## 3.3 The Normal Distribution

Properties of the normal frequency distribution of (3.2) are discussed, for example, in [22] where it is shown that

(i) $f(x)$ is a maximum at $x = \xi$ ($\xi$ being the MPV)
(ii) the curve is symmetrical about $x = \xi$
(iii) points of inflexion occur at $x = \xi \mp \sigma_x$

(iv) approximately 68·3% of the area under the curve lies between
$\xi \pm \sigma_x$
approximately 95·4% of the area under the curve lies between
$\xi \pm 2\sigma_x$
approximately 99·7% of the area under the curve lies between
$\xi \pm 3\sigma_x$

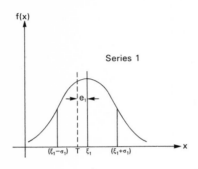 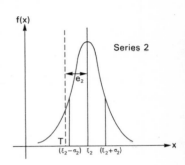

Fig 3.1

In Fig 3.1, two distributions which correspond to two series of repeated measurements of the same quantity are illustrated. Series 1 has expectation $\xi_1$, and variance $\sigma_1^2$, and series 2 has expectation $\xi_2$ and variance $\sigma_2^2$. The second series is more precise than the first ($\sigma_2 < \sigma_1$). Suppose, however, that the true value of the quantity is $x = T$. The accuracy of the first series is therefore indicated by $e_1^{-1}$ and that of series 2, by $e_2^{-1}$. Thus series 2 is less accurate (but more precise) than series 1. $T$ is never known in practice, so the accuracy is indeterminate. Also, in practice, $\xi$ has to be estimated by calculating $\bar{x}$, the mean. The difference between $\bar{x}$ and $\xi$ can be reduced by increasing indefinitely the number of measurements and the difference between $\xi$ and $T$ can be reduced by removing systematic errors and blunders.

## 3.4   Weights

In practice, measurements of differing reliability are often made and it is necessary to find the MPV of the unknown(s) given these measurements with different standard errors. For example, in a triangulation network some angles may have been measured more times (or by a better observer) than others. Again, if both angles and distances have been measured, their relative reliabilities should be taken into account if the MPVs of the coordinates are required.

Suppose that two measurements $x_1$ and $x_2$ have been made and that their respective variances are $\sigma_1^2$ and $\sigma_2^2$. Suppose that the expectation ($\xi$) of each measurement is the same. In order to find the MPV of the quantity measured, consider

$$\tilde{x} = g_1 x_1 + g_2 x_2 \tag{3.3}$$

where $g_1$ and $g_2$ have to be determined so that

(i)  the expectation of $\tilde{x}$ is $\xi$, and
(ii) the variance of $\tilde{x}$ is a minimum (see section 3.2).

From condition (i) above,

$$E\{\tilde{x}\} = \xi = E\{g_1 x_1 + g_2 x_2\} = \xi(g_1 + g_2)$$

(using equations 2.8 and 2.9). Therefore,

$$g_1 + g_2 = 1 \tag{3.4}$$

From condition (ii) above,

$$V\{\tilde{x}\} = V\{g_1 x_1 + g_2 x_2\} = g_1^2 \sigma_1^2 + g_2^2 \sigma_2^2 \tag{3.5}$$

(using equation 2.36 and assuming that $x_1$ and $x_2$ are uncorrelated).
Substituting the value of $g_2$ from (3.4) in (3.5) we have

$$V\{\tilde{x}\} = g_1^2 \sigma_1^2 + (1 - g_1)^2 \sigma_2^2 \tag{3.6}$$

Differentiation of (3.6) with respect to $g_1$ and equation of the derivative to zero gives

$$2 g_1 \sigma_1^2 - 2(1 - g_1)\sigma_2^2 = 0$$

Therefore,

$$g_1 \sigma_1^2 = g_2 \sigma_2^2 \tag{3.7}$$

and

$$\frac{g_1}{g_2} = \frac{\sigma_2^2}{\sigma_1^2} = \frac{w_1}{w_2} \tag{3.8}$$

say, where $w_1$ and $w_2$ are defined as the *weights* of $x_1$ and $x_2$ respectively, which are inversely proportional to the variances of $x_1$ and $x_2$. Therefore, from (3.8) and (3.4)

$$g_1 = w_1/(w_1 + w_2) \quad \text{and} \quad g_2 = w_2/(w_1 + w_2) \tag{3.9}$$

Substitution of these values in (3.3) gives the *weighted mean*

$$\tilde{x} = \frac{w_1 x_1 + w_2 x_2}{w_1 + w_2}$$

This can be proved similarly for any finite number of independent measures $x_i$ with variances $\sigma_i^2$ ($i = 1, 2,..., n$) so that

$$w_1 : w_2 : w_3 : \cdots : w_n = \sigma_1^{-2} : \sigma_2^{-2} : \cdots : \sigma_n^{-2}$$

and the MPV ($\tilde{x}$) of the unknown is given by

$$\tilde{x} = [w_i x_i]/[w_i] \tag{3.10}$$

summation being over $i = 1, 2,..., n$. If we put

$$w_i \sigma_i^2 = \sigma_0^2 \tag{3.11}$$

then $\sigma_0^2$ can be called the *variance of a measurement of unit weight* and (3.5) becomes

$$V\{\tilde{x}\} = \frac{w_1^2 \sigma_1^2 + w_2^2 \sigma_2^2}{(w_1 + w_2)^2} = \frac{w_1 \sigma_0^2 + w_2 \sigma_0^2}{(w_1 + w_2)^2} = \frac{\sigma_0^2}{w_1 + w_2}$$

and, in general terms, when measures $x_i$ of respective weights $w_i$ are made,

$$V\{\tilde{x}\} = \sigma_0^2/[w_i] \tag{3.12}$$

where $\tilde{x}$ is given by (3.10). The dimensions of a weight are those of the reciprocal of the variance. $\sigma_0^2$ is dimensionless. If the weight of a measurement is defined as the reciprocal of the variance of the measurement, then $\sigma_0^2 = 1$. An unbiased estimate of $\sigma_0^2$ can be obtained from the observations by a method similar to that already used to find an unbiased estimate of the variance from the observations (section 2.10). This estimate is given by

$$s_0^2 = [w_i(x_i - \tilde{x})^2]/(n - 1) \tag{3.13}$$

where $E\{s_0^2\} = \sigma_0^2$. Thus, the unbiased estimate of the variance of an observation $x_i$ of weight $w_i$ is, from (3.11), given by $s_{x_i}^2$ where

$$s_{x_i}^2 = \frac{s_0^2}{w_i} = [w_i(x_i - \tilde{x})^2]/w_i(n - 1) \tag{3.14}$$

$s_{x_i}$ is often called the *standard error of an observation of weight $w_i$*. Similarly, the unbiased estimate of the variance of the weighted mean

($\tilde{x}$) is, from (3.12) and (3.13), given by $s_{\tilde{x}}^2$ where

$$s_{\tilde{x}}^2 = s_0^2/[w_i] = [w_i(x_i - \tilde{x})^2]/[w_i](n - 1) \qquad (3.15)$$

$s_{\tilde{x}}$ is often called the *standard error of the weighted mean*.

## 3.5   The Multivariate Normal Distribution

For two independent random variables $x_1$ and $x_2$ with expectations $\xi_1$ and $\xi_2$ and variances $\sigma_1^2$ and $\sigma_2^2$ respectively, the PD surface is

$$f(x_1, x_2) = (2\pi\sigma_1\sigma_2)^{-1} \exp -\tfrac{1}{2}[(x_1 - \xi_1)^2/\sigma_1^2 + (x_2 - \xi_2)^2/\sigma_2^2]$$

$$(3.16)$$

This follows from (3.2) and (2.21). When dealing with $n$-dimensional distributions, it is more convenient (and more illuminating) to write the expression in matrix form. Thus (3.16) becomes

$$f(x_1, x_2) = (2\pi|\sigma_{xx}|^{1/2})^{-1} \exp -\tfrac{1}{2}(\mathbf{x} - \boldsymbol{\xi})^\top \sigma_{xx}^{-1}(\mathbf{x} - \boldsymbol{\xi})$$

where $\mathbf{x} = (x_1\ x_2)^\top$, $\boldsymbol{\xi} = (\xi_1\ \xi_2)^\top$ and

$$\sigma_{xx} = \begin{pmatrix} \sigma_1^2 & 0 \\ 0 & \sigma_2^2 \end{pmatrix}$$

In general, for $n$ independent random variables

$$f(x_1, x_2) = ((2\pi)^{n/2}|\sigma_{xx}|^{1/2})^{-1} \exp -\tfrac{1}{2}(\mathbf{x} - \boldsymbol{\xi})^\top \sigma_{xx}^{-1}(\mathbf{x} - \boldsymbol{\xi})$$

$$(3.17)$$

where $\sigma_{xx}$ is a $(n \times n)$ diagonal matrix. According to the principle of maximum likelihood, the parameters $\xi_1, \xi_2, ..., \xi_n$ must be chosen so that $f(x_1, x_2,..., x_n)$ is a maximum. This will occur when

$$(\mathbf{x} - \boldsymbol{\xi})^\top \sigma_{xx}^{-1}(\mathbf{x} - \boldsymbol{\xi}) \qquad (3.18)$$

is a minimum. In the present case, where the variables are independent, (3.18) reduces to

$$\frac{v_1^2}{\sigma_1^2} + \frac{v_2^2}{\sigma_2^2} + \cdots + \frac{v_n^2}{\sigma_n^2} = \text{minimum} \qquad (3.19)$$

where $v_i = (x_i - \xi_1)$, and is a residual.

If the observations are correlated, then $\sigma_{xx}$ becomes the *variance–*

*covariance matrix* of the observations:

$$\boldsymbol{\sigma}_{xx} = \begin{pmatrix} \sigma_1^2 & \sigma_{12} & \cdots & \sigma_{1n} \\ \sigma_{21} & \sigma_2^2 & \cdots & \sigma_{2n} \\ \vdots & \vdots & & \vdots \\ \sigma_{n1} & \sigma_{n2} & \cdots & \sigma_n^2 \end{pmatrix}$$

Where $\sigma_{ij}$ $(i \neq j)$ is the covariance of $x_i$ and $x_j$ $(= \sigma_{ji})$. In this correlated case, the application of the principle of maximum likelihood leads to the need to minimise the expression $(\mathbf{x} - \boldsymbol{\xi})^{\top}\boldsymbol{\sigma}_{xx}^{-1}(\mathbf{x} - \boldsymbol{\xi})$, which is a quadratic form. When the observations are correlated, the least squares method becomes the method of *minimum quadratic form*.

For a full treatment of the LS principle based on the multivariate normal distribution, refer to [24].

### 3.6 Redundant Observations

In Fig 3.2, A and B are fixed points with known plane rectangular coordinates and C and D are new points in a control survey, whose coordinates are to be determined. There are therefore four unknowns $(E_D, E_C, N_D, N_C)$ and four measurements are therefore necessary and sufficient for the determination of the unknowns.

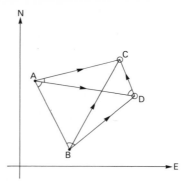

Fig 3.2

These measurements could be

(*a*) angles DAB, DBA, CAD and CDA; or
(*b*) distances AD, BD, AC and DC; or
(*c*) angles ABD, CAD and distances BD, AC;

or any other suitable combination of four measurements.

However, if only four measurements are made, there will be no independent check (section 1.4) and there will be the possibility of an undetected mistake or blunder in the observations. If additional measurements are taken as checks, they can also be used to obtain better estimates of the unknowns. In the framework of Fig 3.2, there are eight interior horizontal angles and five distances (AB is assumed fixed), giving thirteen possible observations to determine four unknowns, i.e. nine redundant observations. Whether or not all thirteen observations are made is a matter to be decided for each individual project; improved estimates of the unknowns must be balanced against the time taken to make the additional observations. Also, the computation necessary if many redundant observations are incorporated could take a long time, if automatic electronic computing facilities are not available. Other factors which will affect the optimum number of observations are the shape of the framework and the specified precision of the coordinates.

The MPVs of the unknowns can be found from the observations by either the so-called *direct method* or the *indirect method*. Both methods make use of the LS principle and both can be applied to the same observations in order to obtain the same unknowns. The latter method (that of *observation equations*) has received increasing attention in recent years, owing in part to the use of digital computers. The former method (that of *condition equations*) has been used extensively in the past and is described with many applications in [22]. The next two sections contain outlines of the theory of each method and a comparison is given in section 3.6.3.

### 3.6.1 The Indirect Method (Observation Equations)
Suppose a measured value of a quantity is $M_x$ and the (unknown) MPV of that quantity is $X$. The residual of the measurement is defined as

$$v = X - M_x \qquad (3.20)$$

This is an *observation equation*. If the quantity measured ($X$) is a linear function of other (unknown) quantities $x_1, x_2,..., x_n$ which are required, then $M$ is an indirect measurement of those quantities and the observation equation becomes

$$v = a_1x_1 + a_2x_2 + \cdots + a_nx_n - M \qquad (3.21)$$

where $a_i$ ($i = 1...n$) is a constant. For $m$ observations ($m > n$) there will be $m$ observation equations such as (3.21):

$$v_1 = a_{11}x_1 + a_{12}x_2 + \cdots + a_{1n}x_n - M_1$$
$$v_2 = a_{21}x_1 + a_{22}x_2 + \cdots + a_{2n}x_n - M_2$$
$$\vdots \qquad \vdots \qquad \vdots \qquad \qquad \vdots \qquad \vdots$$
$$v_m = a_{m1}x_1 + a_{m2}x_2 + \cdots + a_{mn}x_n - M_m$$

or, in matrix notation,

$$\mathbf{v} = \mathbf{Ax} - \mathbf{k} \tag{3.22}$$

where $\mathbf{A}$ is the $m \times n$ matrix of (known) coefficients, $\mathbf{x}$ is the $n \times 1$ matrix of unknowns, $\mathbf{v}$ is the $m \times 1$ matrix of (unknown) residuals and $\mathbf{k}$ is the $m \times 1$ matrix of constant terms, derived from the measurements.

If the measurements are uncorrelated, the LS solution for $\mathbf{x}$ is obtained from (3.22) by minimising $\mathbf{v}^\top \boldsymbol{\sigma}_{kk}^{-1} \mathbf{v}$ where $\boldsymbol{\sigma}_{kk}$ is the $m \times m$ (diagonal) variance–covariance matrix of the measurements (section 3.5). But, in practice, the variances of the measurements are often unknown. At best, we have unbiased estimates—the squares of the standard errors from (2.52) and (2.53)—and these are often based on a small number of measurements. However, in some cases, although the variances are unknown, their ratios are known (see, for example, section 4.3) so we can use a matrix $\mathbf{W}$ instead of $\boldsymbol{\sigma}_{kk}^{-1}$ where

$$\mathbf{W} = \begin{pmatrix} w_1 & 0 & \cdots & 0 \\ 0 & w_2 & \cdots & 0 \\ \vdots & \vdots & & \vdots \\ 0 & 0 & \cdots & w_m \end{pmatrix} = \sigma_0^2 \boldsymbol{\sigma}_{kk}^{-1}$$

and $w_i = \sigma_0^2/\sigma_i^2$, from (3.11).

$\mathbf{W}$ is generally called the *weight matrix* of the (uncorrelated) measurements. Therefore, on the assumption that the ratios of the variances of the observations are accurately known, the LS solution for $\mathbf{x}$ in (3.22) is obtained by minimising $\mathbf{v}^\top \mathbf{W v}$, i.e. by minimising

$$w_1 v_1^2 + w_2 v_2^2 + \cdots + w_m v_m^2$$

The LS solution of $\mathbf{Ax} = \mathbf{v} + \mathbf{k}$ is given by the *normal equations*

$$\mathbf{A}^\top \mathbf{WAx} = \mathbf{A}^\top \mathbf{Wk} \tag{3.23}$$

and it can be written as

$$\mathbf{x} = (\mathbf{A}^\top \mathbf{WA})^{-1} \mathbf{A}^\top \mathbf{Wk} \tag{3.24}$$

The proof of this can be found in [8], for example. It can be seen from (3.24) that the same solution is obtained for **x** whether **W** or $c$**W** is used, where $c$ is an arbitrary constant.

Of particular interest in the solution for **x** is the matrix $(\mathbf{A}^\top \mathbf{WA})^{-1}$, for the variance–covariance matrix $(\mathbf{\sigma}_{xx})$ of the unknowns is shown in [9], for example, to be given by

$$\mathbf{\sigma}_{xx} = \sigma_0^2 (\mathbf{A}^\top \mathbf{WA})^{-1} = \begin{pmatrix} \sigma_{x_1}^2 & \sigma_{x_1 x_2} & \cdots & \sigma_{x_1 x_n} \\ \sigma_{x_2 x_1} & \sigma_{x_2}^2 & \cdots & \sigma_{x_2 x_n} \\ \vdots & \vdots & & \vdots \\ \sigma_{x_n x_1} & \sigma_{x_n x_2} & \cdots & \sigma_{x_n}^2 \end{pmatrix} \tag{3.25}$$

It is extremely useful to have an indication of the precision of the unknowns, and $\mathbf{\sigma}_{xx}$ gives this, provided that a value for $\sigma_0^2$ can be obtained. It is shown in [9] that an unbiased estimator for $\sigma_0^2$ is

$$s_0^2 = \frac{\mathbf{v}^\top \mathbf{W} \mathbf{v}}{m - n} \tag{3.26}$$

where $E\{s_0^2\} = \sigma_0^2$. Having, therefore, obtained **x** from (3.24), this is substituted in the original observation equation (3.22) to obtain **v**. This, in turn, substituted in (3.26), gives the unbiased estimate of $\sigma_0^2$ which enables $\mathbf{\sigma}_{xx}$ to be computed.

If, in the assessment of the elements $w_i$ of the weight matrix **W**, the values chosen were in fact the reciprocals of the variances of the measurements, then $\sigma_0^2$ should equal unity—see (3.11). When $s_0^2$ is computed from (3.26) and its value is different from unity, this indicates that the elements of **W** were not the reciprocals of the variances. Provided, however, that $w_i/w_j = \sigma_j^2/\sigma_i^2$, the LS solution for **x** from (3.24) will be correct. The fact that $E\{s_0^2\} = 1$ can be used to determine the correct ratio of the weights of dissimilar quantities (such as lengths and angles)—see [9], for example.

The relationship expressed by (3.25) enables a control survey to be assessed *before* observations take place. The elements of **A** depend on the geometry of the framework, and if the weights of the observations are assumed, the precision of the coordinates can be determined. It can then be seen where the framework is weak and the geometry or the weights of the observations can be adjusted to improve the precision of the coordinates. All this can be done before observation starts.

A numerical example of the application of the method of observation equations is given in section 4.4.

### 3.6.2 *The Direct Method (Condition Equations)*
Suppose that the MPVs of $n$ observed quantities $(X_1, X_2,..., X_n)$ have to satisfy $m$ *independent* linear condition equations $(m < n)$, i.e.

$$a_{11}X_1 + a_{12}X_2 + \cdots + a_{1n}X_n = L_1$$
$$a_{21}X_1 + a_{22}X_2 + \cdots + a_{2n}X_n = L_2$$
$$\vdots \quad\quad \vdots \quad\quad\quad\quad \vdots \quad\quad \vdots$$
$$a_{m1}X_1 + a_{m2}X_2 + \cdots + a_{mn}X_n = L_m$$

Substituting for $X_i = v_i + M_i$ from (3.20) we have

$$a_{11}v_1 + a_{12}v_2 + \cdots + a_{1n}v_n = L_1 - (a_{11}M_1 + a_{12}M_2 + \cdots + a_{1n}M_n)$$
$$a_{21}v_1 + a_{22}v_2 + \cdots + a_{2n}v_n = L_2 - (a_{21}M_1 + a_{22}M_2 + \cdots + a_{2n}M_n)$$
$$\vdots \quad\quad \vdots \quad\quad\quad \vdots \quad\quad \vdots \quad\quad \vdots \quad\quad\quad \vdots$$
$$a_{m1}v_1 + a_{m2}v_2 + \cdots + a_{mn}v_n = L_m - (a_{m1}M_1 + a_{m2}M_2 + \cdots + a_{mn}M_n)$$

or $$\mathbf{Av} = \mathbf{b} \tag{3.27}$$

The LS solution is obtained by minimising $\mathbf{v}^\top \mathbf{Wv}$, which is the same as minimising $\mathbf{v}^\top \mathbf{Wv} - 2(\mathbf{Av} - \mathbf{b})^\top \mathbf{k}$, where $\mathbf{k}$ is the $m \times 1$ matrix of *Lagrange coefficients* or *correlates*. It is shown in [8] that this, coupled with the requirement to fulfil (3.27), leads to the relation

$$\mathbf{v} = \mathbf{W}^{-1}\mathbf{A}^\top\mathbf{k} \tag{3.28}$$

where $\mathbf{k}$ is obtained from the *correlative normals*

$$(\mathbf{AW}^{-1}\mathbf{A}^\top)\mathbf{k} = \mathbf{b} \tag{3.29}$$

Thus, substitution of $\mathbf{k}$ from (3.29) in (3.28) gives the $n \times 1$ matrix $\mathbf{v}$ as

$$\mathbf{v} = \mathbf{W}^{-1}\mathbf{A}^\top(\mathbf{AW}^{-1}\mathbf{A}^\top)^{-1}\mathbf{b} \tag{3.30}$$

This solution for $v_1, v_2,..., v_n$ enables the MPVs $(X_1, X_2,..., X_n)$ of the measured quantities to be found from $X_i = M_i + v_i$. As in the indirect method of the preceding section, the variance–covariance matrix of the unknowns can be determined, although here it is more complex. The relevant expression is derived in [24], where the unbiased estimator for $\sigma_0^2$ is also given.

Numerical examples of the application of this method are given in section 4.5.

### 3.6.3 *The Methods Compared*

One important disadvantage of the condition equations method is that it is essential to determine the conditions necessary and sufficient for a unique solution, and this is often very difficult. Various methods are available to ensure that the correct number of conditions are arrived at and these are discussed in [22]. On the other hand, it is not necessary to consider the geometrical conditions when using the observation equations method: for each observed quantity, there will be one equation.

When using observation equations, there is not as good a check on mistakes in the observations as there is in the condition equations method. The right-hand side of (3.27) is an indication of the amount by which the measurements fail to fulfil the conditions; any large term appearing here indicates a probable mistake in the measurements.

Both methods involve the solution of the equations of the form $\mathbf{B}^\mathsf{T}\mathbf{B}\mathbf{x} = \mathbf{c}$ (the normal equations) where the matrix $\mathbf{B}^\mathsf{T}\mathbf{B}$ is symmetric and positive–definite. This property is important as it leads to more rapid methods of solution. The smaller the matrix, the quicker the solution and the less storage space needed for a given method of solution. In the case of observation equations, the coefficient matrix of the normals will have the dimensions of the number of independent unknowns, whereas in the case of condition equations it will have the dimensions of the number of independent conditions. The dimensions of this matrix will probably be the factor which determines the method to be used when the solution is by calculators with few stores. Observation equations should be used for most problems in modern engineering practice, if the computing facilities are available, and condition equations for simple triangulation figures and traverses when there are only limited computing facilities—i.e. hand calculators and desk computers.

# CHAPTER FOUR

# The Application of Statistical Principles to Surveying

## 4.1   The MPV and Precision of a Series of Measurements

Suppose ten equally reliable measurements of an angle have been made with values as given in Fig 4.1.

| Measured value ($\alpha$) | Residual ($v''$) | $v^2$ |
|---|---|---|
| 4° 27′ 20″ | − 3·7 | 13·69 |
| 14 | + 2·3 | 5·29 |
| 17 | − 0·7 | 0·49 |
| 14 | + 2·3 | 5·29 |
| 16 | + 0·3 | 0·09 |
| 21 | − 4·7 | 22·09 |
| 12 | + 4·3 | 18·49 |
| 16 | + 0·3 | 0·09 |
| 18 | − 1·7 | 2·89 |
| 15 | + 1·3 | 1·69 |

Fig 4.1

According to the LS principle of Legendre (section 3.1) the arithmetic mean is the MPV since the observations are equally reliable. This mean ($\bar{\alpha}$) is 4° 27′ 16″·3. The precision of a measured value is given by (2.52) as the standard error, $s_\alpha$ ;

$$s_\alpha = \sqrt{\frac{[v^2]}{n-1}} = \sqrt{\frac{70·10}{10-1}} \simeq \pm 2''·79$$

According to (2.53) the standard error of the mean is

$$s_{\bar{a}} = s_\alpha/\sqrt{n} = \pm 2\cdot 79/\sqrt{10} \simeq \pm 0''\cdot 88$$

Thus, the measurement can be quoted as

$$= 4° \ 27' \ 16''\cdot 3 \pm 0''\cdot 88$$

Now suppose that the same observer, using the same equipment and the same procedure, and working under the same conditions, is called upon to measure an angle to give a standard error in the mean of $\pm 1''$. The above results indicate that each angle should be measured $N$ times where $N$ is given by

$$\pm 1 = 2\cdot 79/\sqrt{N}$$

Thus, in this case, $N = 7\cdot 78$, say eight, times. If, however, a precision in the mean of $\pm 2''$ is specified, there need only be two measurements. Increasing the precision $x$ times means increasing the number of measurements $x^2$ times.

## 4.2   The Estimate of Covariance

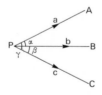

Fig 4.2

In Fig 4.2, P, A, B and C are four points in a survey framework, where the horizontal angles $\alpha$ and $\beta$ are measured. Suppose an observer at P takes a succession of theodolite horizontal circle readings pointing to A, then to B, then to C, then to A again, etc., and obtains the three series of readings $a$, $b$ and $c$ shown in Fig 4.3. The three arithmetic means are 28° 47′ 32″·1, 47° 18′ 19″·4 and 69° 50′ 32″·3. The residuals are shown in Fig 4.4 as $v_a$, $v_b$ and $v_c$ respectively. The standard errors of the measurements of $a$, $b$ and $c$ are given by

$$s_a^2 = [v_a^2]/(n-1) = 146\cdot 9 \div 9 \simeq 16\cdot 32 \ \text{sec}^2$$
$$s_b^2 = [v_b^2]/(n-1) = \ \ 76\cdot 4 \div 9 \simeq \ \ 8\cdot 49 \ \text{sec}^2$$
$$s_c^2 = [v_c^2]/(n-1) = \ \ 36\cdot 1 \div 9 \simeq \ \ 4\cdot 01 \ \text{sec}^2$$

| a | b | c |
|---|---|---|
| 28° 47′ 29″ | 47° 18′ 19″ | 69° 50′ 34″ |
| 34 | 20 | 35 |
| 28 | 18 | 35 |
| 33 | 17 | 33 |
| 35 | 24 | 31 |
| 35 | 18 | 29 |
| 31 | 16 | 30 |
| 29 | 25 | 32 |
| 27 | 19 | 32 |
| 40 | 18 | 32 |

Fig 4.3

| $v_a$ | $v_a^2$ | $v_b$ | $v_b^2$ | $v_c$ | $v_c^2$ | $v_a v_b$ | $v_b v_c$ | $v_a v_c$ |
|---|---|---|---|---|---|---|---|---|
| −3·1 | 9·61 | −0·4 | 0·16 | +1·7 | 2·89 | + 1·24 | −0·68 | − 5·27 |
| +1·9 | 3·61 | +0·6 | 0·36 | +2·7 | 7·29 | + 1·14 | +1·62 | + 5·13 |
| −4·1 | 16·81 | −1·4 | 1·96 | +2·7 | 7·29 | + 5·74 | −3·78 | −11·07 |
| +0·9 | 0·81 | −2·4 | 5·76 | +0·7 | 0·49 | − 2·16 | −1·68 | + 0·63 |
| +2·9 | 8·41 | +4·6 | 21·16 | −1·3 | 1·69 | +13·34 | −5·98 | − 3·77 |
| +2·9 | 8·41 | −1·4 | 1·96 | −3·3 | 10·89 | − 4·06 | +4·62 | − 9·57 |
| −1·1 | 1·21 | −3·4 | 11·56 | −2·3 | 5·29 | + 3·74 | +7·82 | + 2·53 |
| −3·1 | 9·61 | +5·6 | 31·36 | −0·3 | 0·09 | −17·36 | −1·68 | + 0·93 |
| −5·1 | 26·01 | −0·4 | 0·16 | −0·3 | 0·09 | + 2·04 | +0·12 | + 1·53 |
| +7·9 | 62·41 | −1·4 | 1·96 | −0·3 | 0·09 | −11·06 | +0·42 | − 2·37 |

Fig 4.4

These values indicate that the pointings to C are more precise than those to B, which in turn are more precise than those to A (probably owing to the different types of targets used, since the distances are roughly equal). The covariance can be estimated by

$$S_{ab} = [v_a v_b]/(n-1) = -7.4 \div 9 \simeq -0.82 \text{ sec}^2$$
$$S_{bc} = [v_b v_c]/(n-1) = +0.8 \div 9 \simeq +0.09 \text{ sec}^2$$
$$S_{ac} = [v_a v_c]/(n-1) = -21.3 \div 9 \simeq -2.37 \text{ sec}^2$$

Therefore the coefficients of correlation are assessed as

$$\rho_{ab} = S_{ab}/S_a S_b = -0.82/(4.04 \times 2.91) \simeq -0.07$$
$$\rho_{bc} = S_{bc}/S_b S_c = +0.09/(2.91 \times 2.00) \simeq +0.02$$
$$\rho_{ac} = S_{ac}/S_a S_c = -2.37/(4.04 \times 2.00) \simeq -0.29$$

This indicates that there is very little correlation between the pointings.

The angles are given by $\alpha = (b - a)$ and $\beta = (c - b)$. Applications of (2.26) give

$$s_\alpha^2 = s_a^2 + s_b^2 - 2s_{ab} \simeq 16.32 + 8.49 - 2(-0.82) \simeq 26.45 \text{ sec}^2$$
$$s_\beta^2 = s_b^2 + s_c^2 - 2s_{bc} \simeq 8.49 + 4.01 - 2(+0.09) \simeq 12.32 \text{ sec}^2$$

and the covariance of $\alpha$ and $\beta$ is given by (2.57) as

$$
S_{\alpha\beta} = \left(\frac{\partial\alpha}{\partial a}\cdot\frac{\partial\beta}{\partial a}\right)\cdot s_a^2 + \left(\frac{\partial\alpha}{\partial b}\cdot\frac{\partial\beta}{\partial b}\right)\cdot s_b^2 + \left(\frac{\partial\alpha}{\partial c}\cdot\frac{\partial\beta}{\partial c}\right)\cdot s_c^2
$$
$$
+ \left(\frac{\partial\alpha}{\partial a}\cdot\frac{\partial\beta}{\partial b} + \frac{\partial\alpha}{\partial b}\cdot\frac{\partial\beta}{\partial a}\right)\cdot s_{ab} + \left(\frac{\partial\alpha}{\partial a}\cdot\frac{\partial\beta}{\partial c} + \frac{\partial\alpha}{\partial c}\cdot\frac{\partial\beta}{\partial a}\right)\cdot s_{ac}
$$
$$
+ \left(\frac{\partial\alpha}{\partial b}\cdot\frac{\partial\beta}{\partial c} + \frac{\partial\alpha}{\partial c}\cdot\frac{\partial\beta}{\partial b}\right)\cdot s_{bc}
$$
$$
= 0.s_a^2 - 1.s_b^2 + 0.s_c^2 + (1+0).s_{ab} + (-1+0).s_{ac}
$$
$$
+ (1+0).s_{bc}
$$
$$
\simeq -8.49 - 0.82 + 2.37 + 0.09 = -6.85 \text{ sec}^2
$$

Thus $\quad \rho_{\alpha\beta} = S_{\alpha\beta}/S_\alpha S_\beta = -6.85/(5.14 \times 3.51) \simeq -0.38 \text{ sec}^2$

(Note that $\gamma = (c - a)$, so that $s_\gamma^2 = s_a^2 + s_c^2 - 2s_{ac} = 25.07 \text{ sec}^2$, and since $\gamma = \alpha + \beta$, $s_\gamma^2 = s_\alpha^2 + s_\beta^2 + 2s_{\alpha\beta} = 25.07 \text{ sec}^2$ also.)

In practice, it is generally not possible to calculate estimates of the correlation between observed directions, as in the above example. However, it should be possible to deduce or assume reasonable values for the standard errors of the observed directions. This then enables estimates of the variances of the angles to be made, using, for example, $s_\alpha^2 = s_a^2 + s_b^2$. Also, estimates of the covariance of angles at a point can be made as shown above, with $s_{ab} = s_{ac} = s_{bc} = 0$. Then $s_{\alpha\beta} = -s_b^2$.

## 4.3   The Propagation of Random Errors

In this section various common survey procedures are considered, and the way in which random errors are propagated during each procedure is discussed. This enables a weighting system to be devised.

### 4.3.1   *The Propagation of Random Errors in Levelling*
Suppose that a line of levels is run from A to B, a total distance $D$. Suppose that at every set up of the level, the backsight and foresight distances are the same and equal to $d$. There will then be $n$ level positions (where $n = D/2d$); Fig 4.5 shows such a level position. Suppose

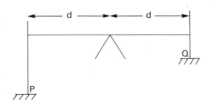

Fig 4.5

each staff reading has a standard error of $\pm s$. Thus, if $R_p$ and $R_q$ are the (uncorrelated) staff readings at P and Q respectively, the 'observed' height difference between P and Q is

$$h = R_p - R_q \tag{4.1}$$

and according to (2.56) and (4.1) the standard error in $h$ ($s_h$) is given by

$$s_h^2 = s^2 + s^2 = 2s^2 \tag{4.2}$$

There will be $n$ such typical height differences between A and B, so the standard error of the height difference between A and B ($s_H$) is given by

$$s_H^2 = n.s_h^2 = 2ns^2 = (Ds^2)/d \tag{4.3}$$

that is
$$s_H \propto \sqrt{D} \qquad (4.4)$$
(compare this with equation 4.7).

For a given sighting distance $(d)$ and standard error of staff reading $(s)$ the standard error of the levelling can be computed from (4.3). Conversely, if the standard error $s_H$ is specified, tests to determine the variation of $s$ with $d$ can be made to determine the best observing procedure to be used to meet the specification.

### 4.3.2  The Propagation of Random Errors in Trig. Heighting
Suppose that the vertical angle $\theta$ is observed from A to B shown in Fig 4.6 in order to find the height of B $(H_B)$.

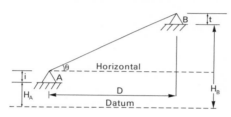

Fig 4.6

This is given by
$$H_B = H_A + i - t + D \tan \theta \qquad (4.5)$$
If the standard errors of $H_A$, $i$, $t$ and $D$ are all assumed to be zero, then the standard error of $H_B$ $(s_H)$ arising from a standard error of $s_\theta$ in measuring $\theta$ is, from (2.56) and (4.5), given by
$$s_H^2 = D^2 \sec^4 \theta . s_\theta^2 \qquad (4.6)$$
and therefore
$$s_H \propto D \qquad (4.7)$$

The limitations of such a conclusion should be mentioned here, since probably the greatest error in trig. heighting will arise not from the random errors but from systematic effects of refraction and earth curvature. The simple mathematical model of (4.5) is generally inadequate except for an investigation of the effects of random errors in the observations, such as is given here. For a discussion of the effects of earth curvature and atmospheric refraction, see section 6.5.

### 4.3.3  The Propagation of Random Errors in Plane Fixation
In Fig 4.7, A and B are points with known coordinates, $(E_A, N_A)$ and $(E_B, N_B)$ respectively, and therefore the bearing $\alpha$ is also known.

Measurements of the horizontal angle BAP $= \theta$ and the horizontal distance AP $= L$ enable the coordinates of P $(E_P, N_P)$ to be calculated from

$$E_P = E_A + L \sin(\alpha + \theta) \tag{4.8}$$

and $$N_P = N_A + L \cos(\alpha + \theta) \tag{4.9}$$

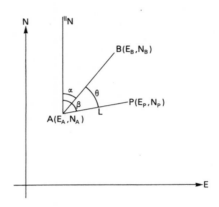

Fig 4.7

The inevitable presence of random errors in the measurements of $L$ and $\theta$ will result in random errors in $E_P$ and $N_P$. An estimate of these indicates the precision with which P has been fixed and it can be seen whether a specification has been met. In order to simplify this example, assume that $E_A$, $N_A$, $E_B$ and $N_B$ are error-free and that $\theta$ and $L$ are uncorrelated. Equations (2.56), (4.8) and (4.9) give the standard errors of $E_P$ and $N_P$ ($s_E$ and $s_N$ respectively) from

$$s_E^2 = s_L^2 \sin^2 \beta + L^2 s_\theta^2 \cos^2 \beta \tag{4.10}$$

and $$s_N^2 = s_L^2 \cos^2 \beta + L^2 s_\theta^2 \sin^2 \beta \tag{4.11}$$

noting that $\beta = (\alpha + \theta)$, so $s_\beta^2 = s_\theta^2$, since $s_\alpha^2 = 0$. The covariance estimate $(s_{EN})$ of $E_P$ and $N_P$ is, from (2.57), (4.8) and (4.9)

$$s_{EN} = s_L^2 \sin \beta \cos \beta - L^2 s_\theta^2 \sin \beta \cos \beta$$
$$= \tfrac{1}{2}(s_L^2 - L^2 s_\theta^2) \sin 2\beta \tag{4.12}$$

Equations (4.10) and (4.11) give the standard errors of the position of P in the N–S and E–W directions only. The standard error in any direction $\phi$ from north is of greater importance. To obtain this general expression as a function of $\phi$, consider a rotation of the axes through an angle $\phi$ such that $(E, N)$ becomes $(e, n)$. According to (5.5)

$$n = N \cos \phi + E \sin \phi \qquad (4.13)$$

Therefore the standard error $(s_n)$ in the $n$-direction is, from (2.56) and (4.13), given by

$$s_n^2 = s_N^2 \cos^2 \phi + s_E^2 \sin^2 \phi + 2s_{EN} \sin \phi \cos \phi \qquad (4.14)$$

For any given fixation $s_N^2$, $s_E^2$ and $s_{EN}$ are constant and (4.14) represents the polar equation of the pedal curve of the error ellipse of the fixation (see equation 4.20 and its derivation). Maximum and minimum values of $s_n$ occur at values of $\phi$ given by

$$(\mathrm{d}/\mathrm{d}\phi)(s_n^2) = 2 \sin \phi \cos \phi(s_E^2 - s_N^2) + 2s_{EN}(\cos^2 \phi - \sin^2 \phi) = 0$$

i.e. when $\qquad \phi = \tfrac{1}{2} \tan^{-1}[2s_{EN}/(s_N^2 - s_E^2)] \qquad (4.15)$

As a numerical illustration, consider the fixation of P from A and B, where

$$E_A = 1760 \cdot 39 \text{ m} \qquad L = 917 \cdot 29 \pm 0 \cdot 07 \text{ m}$$
$$N_A = 4830 \cdot 51 \text{ m} \qquad \theta = 42° \ 18' \ 40'' \pm 8''$$
$$\alpha = 18° \ 28' \ 17'' \qquad \beta = (\alpha + \theta) = 60° 46' 57'' \pm 8''$$

Thus,

$$E_P = E_A + L \sin \beta = 1760 \cdot 39 + 917 \cdot 29 \times 0 \cdot 872 \ 773 = 2560 \cdot 98 \text{ m}$$
$$N_P = N_A + L \cos \beta = 4830 \cdot 51 + 917 \cdot 29 \times 0 \cdot 488 \ 126 = 5278 \cdot 26 \text{ m}$$

Equations (4.10), (4.11) and (4.12) give, respectively,

$$s_E^2 = (0 \cdot 07)^2(0 \cdot 8728)^2 + (917)^2 \left(\frac{8}{2 \cdot 063 \times 10^5}\right)^2 (0 \cdot 4881)^2$$

$$\simeq 4030 \text{ mm}^2$$

$$s_N^2 = (0 \cdot 07)^2(0 \cdot 4881)^2 + (917)^2 \left(\frac{8}{2 \cdot 063 \times 10^5}\right)^2 (0 \cdot 8728)^2$$

$$\simeq 2130 \text{ mm}^2$$

$$s_{EN} = (0 \cdot 8728)(0 \cdot 4881)\left[(0 \cdot 07)^2 - (917)^2 \left(\frac{8}{2 \cdot 063 \times 10^5}\right)^2\right]$$

$$\simeq 1550 \text{ mm}^2$$

Thus, the polar equation of the pedal curve for this fixation is, from

(4.14), given by

$$s_n^2 = 10^4(0{\cdot}213 \cos^2 \phi + 0{\cdot}403 \sin^2 \phi + 0{\cdot}310 \sin \phi \cos \phi)$$

This curve is illustrated in Fig. 4.8. Maxima occur at $\phi = \beta$ and at $\phi = \beta + \pi$, when $s_n = s_L = 70$ mm. Minima occur at $\phi = \beta + \pi/2$ and at $\phi = \beta + 3\pi/2$, when $s_n = Ls_\theta = 35{\cdot}6$ mm.

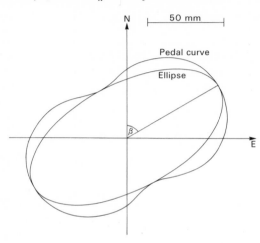

Fig 4.8

### 4.3.4 The Error Ellipse
Suppose the equation of the ellipse $(e)$ shown in Fig 4.9 is

$$ay^2 + 2cxy + bx^2 = 1 \qquad (4.16)$$

The gradient at any point is given by

$$\frac{dy}{dx} = -\left(\frac{bx + cy}{ay + cx}\right)$$

and thus the equation of the tangent $(t)$ at A $(x_A, y_A)$ is

$$y - y_A = -\left(\frac{bx_A + cy_A}{ay_A + cx_A}\right)(x - x_A)$$

or
$$y_A(ay + cx) + x_A(bx + cy) = 1 \qquad (4.17)$$

The equation of the line $(L)$ through O, perpendicular to $t$, is

$$y = \left(\frac{ay_A + cx_A}{bx_A + cy_A}\right)x \qquad (4.18)$$

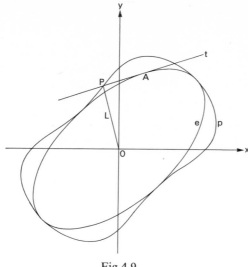

Fig 4.9

The lines $t$ and $L$ intersect at P. As A moves along the ellipse, the locus of P is the pedal curve ($p$) of the given ellipse. Its equation can be found by solving (4.17) and (4.18) for $(x_A, y_A)$ and by substitution of these values in (4.16). Thus, the equation of the pedal curve is

$$ax^2 - 2cxy + by^2 = (x^2 + y^2)(ab - c^2) \qquad (4.19)$$

To convert this to polar form, substitute $x = r \sin \phi$ and $y = r \cos \phi$ to give

$$r^2(ab - c^2) = a \sin^2 \phi + b \cos^2 \phi - 2c \sin \phi \cos \phi \quad (4.20)$$

## 4.4  Variation of Coordinates

This is the name given to one method of computing the MPVs of $n$ coordinates of points given $m$ angular and linear measurements between them ($m > n$). The coordinates can be spheroidal, three dimensional rectangular, plane rectangular or of any other kind; only plane rectangular $(E, N)$ will be considered here, since this is the simplest case and also that most commonly encountered in engineering control surveys.

The observation equation (3.20) $v = X - M$ is written for each measurement. For a measured distance such as that between $i$ and $j$ in Fig 4.10 the equation becomes

$$v_{ij} = L_{ij} - M_{ij} \qquad (4.21)$$

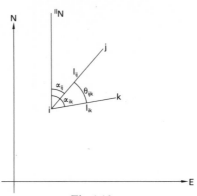

Fig 4.10

where $v_{ij}$ is the residual of the measurement, $L_{ij}$ is the MPV of the distance and $M_{ij}$ is the measured value of the distance. However, since the MPVs of the coordinates of $i$ and of $j$ are required, we can substitute for $L_{ij}$ in (4.21) from

$$L_{ij} = [(E_j - E_i)^2 + (N_j - N_i)^2]^{1/2}$$

where $E_i$, $N_i$, $E_j$ and $N_j$ are the MPVs of the coordinates. This results in a non-linear observation equation, whereas the method of solution (section 3.6.1) depends upon the equations being linear.

In order to obtain a linear observation equation, take approximate coordinates $e_i$, $n_i$, $e_j$ and $n_j$ such that $E_i = e_i + \delta e_i$, etc. If $l_{ij}$ is the approximate distance,

$$l_{ij}^2 = (e_j - e_i)^2 + (n_j - n_i)^2 \qquad (4.22)$$

and

$$L_{ij} = l_{ij} + \delta l_{ij} \qquad (4.23)$$

Differentiation of (4.22) gives

$$2l_{ij}\,\delta l_{ij} = -2(e_j - e_i)\,\delta e_i + 2(e_j - e_i)\,\delta e_j - 2(n_j - n_i)\,\delta n_i \\ + 2(n_j - n_i)\,\delta n_j$$

or

$$\delta l_{ij} = -\delta e_i \sin \alpha_{ij} + \delta e_j \sin \alpha_{ij} - \delta n_i \cos \alpha_{ij} + \delta n_j \cos \alpha_{ij} \qquad (4.24)$$

where $\alpha_{ij}$ is the bearing of line $ij$ computed from the approximate coordinates. Substitution for $L_{ij}$ in (4.21) from (4.23) and (4.24) gives

$$v_{ij} = -\delta e_i \sin \alpha_{ij} + \delta e_j \sin \alpha_{ij} - \delta n_i \cos \alpha_{ij} + \delta n_j \cos \alpha_{ij} - M_{ij}$$
$$+ l_{ij}$$

In this observation equation, the unknowns are $\delta e_i$, $\delta n_i$, $\delta e_j$ and $\delta n_j$. Their coefficients (and $l_{ij}$) are computed from the assumed values $e_i$, $n_i$, $e_j$ and $n_j$. The term $M_{ij}$ is the measured value or 'observed' ($O$) term and $l_{ij}$ is the 'computed' ($C$) term. The usual form of the observation equation is

$$v_{ij} = -\delta e_i \sin \alpha_{ij} + \delta e_j \sin \alpha_{ij} - \delta n_i \cos \alpha_{ij} + \delta n_j \cos \alpha_{ij}$$
$$- (O - C)_{ij} \qquad (4.25)$$

For each observed horizontal angle there will be an observation equation

$$v_{ijk} = \Theta_{ijk} - M_{ijk} \qquad (4.26)$$

where $\Theta_{ijk}$ is the MPV of the angle and $M_{ijk}$ is the observed value. In Fig 4.10, $\theta_{ijk}$ is the horizontal angle at $i$ reckoned clockwise from $j$ to $k$, computed from the approximate coordinates of $i$, $j$ and $k$. $\theta_{ijk}$ is given by

$$\theta_{ijk} = \alpha_{ik} - \alpha_{ij}$$

Therefore
$$\delta\theta_{ijk} = \delta\alpha_{ik} - \delta\alpha_{ij} \qquad (4.27)$$

But
$$\tan \alpha_{ij} = (e_j - e_i)/(n_j - n_i)$$

differentiation and simplification of which gives

$$l_{ij}\,\delta\alpha_{ij} = -\delta e_i \cos \alpha_{ij} + \delta e_j \cos \alpha_{ij} + \delta n_i \sin \alpha_{ij} - \delta n_j \sin \alpha_{ij}$$
$$(4.28)$$

Similarly,

$$l_{ik}\,\delta\alpha_{ik} = -\delta e_i \cos \alpha_{ik} + \delta e_k \cos \alpha_{ik} + \delta n_i \sin \alpha_{ik} - \delta n_k \sin \alpha_{ik}$$
$$(4.29)$$

Substitution for $\Theta_{ijk}$ in (4.26) from (4.27), (4.28) and (4.29) gives

$$v_{ijk} = \delta e_i \left( \frac{\cos \alpha_{ij}}{l_{ij}} - \frac{\cos \alpha_{ik}}{l_{ik}} \right) - \delta e_j \frac{\cos \alpha_{ij}}{l_{ij}} + \delta e_k \frac{\cos \alpha_{ik}}{l_{ik}}$$
$$+ \delta n_i \left( -\frac{\sin \alpha_{ij}}{l_{ij}} + \frac{\sin \alpha_{ik}}{l_{ik}} \right) + \delta n_j \frac{\sin \alpha_{ij}}{l_{ij}} - \delta n_k \frac{\sin \alpha_{ik}}{l_{ik}}$$
$$- (O - C)_{ijk} \qquad (4.30)$$

Thus, for the $m$ observations there will be $m$ equations in $n$ unknowns, the observation equations being as (3.22): $\mathbf{v} = \mathbf{Ax} - \mathbf{k}$ where $\mathbf{v}$ is the $m \times 1$ matrix of (unknown) residuals, $\mathbf{A}$ is the $m \times n$ matrix of coefficients (computed from the approximate coordinates), $\mathbf{x}$ is the $n \times 1$ matrix of corrections to the approximate coordinates and $\mathbf{k}$ is the $m \times 1$ matrix of constant $(O - C)$ terms.

The solution gives the 'corrections' to the approximate coordinates. This enables new values of the coordinates to be found, which are used to re-calculate the coefficients and $C$ terms in the observation equations. Iterative solutions are found until the 'corrections' are insignificant.The number of iterations depends primarily on the accuracy of the assumed 'starting values' of the coordinates and the shape of the network.

If some coordinate values in the network are considered error-free, and are to be held fixed in the LS adjustment, then their 'corrections', $\delta e$ and $\delta n$, are put equal to zero in each observation equation in which they appear.

The numerical example which follows is an almost trivial application insofar as only four observations are used to determine only two unknowns. Such a small problem, however, enables the reader to follow the method step by step in numerical form.

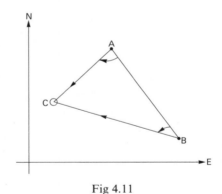

Fig 4.11

In Fig 4.11, A and B are fixed, with error-free coordinates. Point C is located by angles A and B and distances AC and BC.

(a) *Data:* $\quad E_A = 617\ 994{\cdot}28$ m $\qquad\qquad E_B = 618\ 153{\cdot}23$ m

$\qquad\qquad N_A = 136\ 226{\cdot}17$ m $\qquad\qquad N_B = 135\ 872{\cdot}46$ m

$\qquad\qquad$ Bearing AB $= 155°\ 48'\ 05''$

(b) *Measurements (O-values):* $\angle A = 55° \; 51' \; 56'' \pm 3''$

$\angle B = 41° \; 24' \; 49'' \pm 3''$

$AC = 258 \cdot 599 \pm 0 \cdot 003$ m

$BC = 323 \cdot 590 \pm 0 \cdot 003$ m

(c) *C-values:* Approximate coordinates of C are 617 858 m E

136 006 m N

For AC

$$\Delta E = -136 \cdot 28 \text{ m} \qquad \tan \alpha_{AC} = +0 \cdot 618 \; 976$$

$$\Delta N = -220 \cdot 17 \text{ m} \qquad \alpha_{AC} = 211° \; 45' \; 23''$$

$$\left. \begin{array}{l} l_{AC} = \Delta E / \sin \alpha_{AC} = 258 \cdot 935 \text{ m} \\ \phantom{l_{AC}} = \Delta N / \cos \alpha_{AC} = 258 \cdot 934 \text{ m} \end{array} \right\} \text{check}$$

For BC, similarly,

$$\alpha_{BC} = 294° \; 20' \; 18''$$

$$l_{BC} = 324 \cdot 028 \text{ m}$$

For angle BAC

$$211° \; 45' \; 23'' - 155° \; 48' \; 05'' = 55° \; 57' \; 18''$$

For angle ABC

$$335° \; 48' \; 05'' - 294° \; 20' \; 18'' = 41° \; 27' \; 47''$$

(d) *Formation of observation equations:*
(note that $\delta e_A = \delta n_A = \delta e_B = \delta b_B = 0$)

(i) For distance AC

$$v_{AC} = -0 \cdot 526 \; 3 \; \delta e_C - 0 \cdot 850 \; 3 \; \delta n_C - (258 \cdot 599 - 258 \cdot 934) \text{ m}$$

(ii) For distance BC

$$v_{BC} = -0 \cdot 911 \; 1 \; \delta e_C + 0 \cdot 412 \; 1 \; \delta n_C - (323 \cdot 590 - 324 \cdot 028) \text{ m}$$

(iii) For angle BAC

$$v_{ABC} = \frac{-0 \cdot 850 \; 3}{258 \cdot 9} \delta e_C + \frac{0 \cdot 526 \; 3}{258 \cdot 9} \delta n_C - (55° \; 51' \; 56'' - 55° \; 57' \; 18'') \text{ rad}$$

(iv) For angle ABC

$$v_{BCA} = \frac{-0 \cdot 412 \; 1}{324 \cdot 0} \delta e_C - \frac{0 \cdot 911 \; 1}{324 \cdot 0} \delta n_C - (41° \; 24' \; 49'' - 41° \; 27' \; 47'') \text{ rad}$$

Thus the observation equations are $v = Ax - k$, or

$$v = \begin{pmatrix} -0.526\ 3 & -0.850\ 3 \\ -0.911\ 1 & +0.412\ 1 \\ -0.003\ 284 & +0.002\ 033 \\ -0.001\ 272 & -0.002\ 812 \end{pmatrix} \begin{pmatrix} \delta e_C \\ \delta n_C \end{pmatrix} - \begin{pmatrix} -0.335\ 0 \\ -0.438\ 0 \\ -0.001\ 561 \\ -0.000\ 863 \end{pmatrix}$$

(e) *The weight matrix*

$$s_{AC} = s_{BC} = \pm\ 3 \times 10^{-3}\ m$$

Therefore, let

$$w_{AC} = w_{BC} = (\tfrac{1}{9} \times 10^{-6})\ m^{-2} = 0.111\ 1 \times 10^6\ m^{-2}$$

Also,

$$s_{BAC} = s_{ABC} = \pm 3''$$

Therefore, let

$$w_{ABC} = w_{BAC} = (\tfrac{1}{9})\ sec^{-2} = 0.472\ 7 \times 10^{10}\ rad^{-2}$$

Therefore,

$$W = 10^6 \begin{pmatrix} 0.111\ 1 & 0 & 0 & 0 \\ 0 & 0.111\ 1 & 0 & 0 \\ 0 & 0 & 0.472\ 7 \times 10^4 & 0 \\ 0 & 0 & 0 & 0.472\ 7 \times 10^4 \end{pmatrix}$$

(f) *The solution*

This is given as $x = (A^TWA)^{-1}A^TWk$

$$(A^TWA) = 10^6 \begin{pmatrix} +0.181\ 6 & -0.006\ 6 \\ -0.006\ 6 & +0.156\ 1 \end{pmatrix}$$

$$(A^TWA)^{-1} = 10^{-6} \begin{pmatrix} +5.517\ 5 & +0.233\ 5 \\ +0.233\ 5 & +6.415\ 4 \end{pmatrix}$$

$$(A^TWk) = 10^6 \begin{pmatrix} +0.093\ 37 \\ +0.008\ 03 \end{pmatrix}$$

$$x = \begin{pmatrix} +0.517 \\ +0.073 \end{pmatrix} = \begin{pmatrix} \delta e_C \\ \delta n_C \end{pmatrix}$$

Therefore $\quad \delta e_C = +0.52\ m \quad$ and $\quad E_C = 617\ 858.52\ m$

and $\qquad \delta n_C = +0.07\ m \quad$ and $\quad N_C = 136\ 006.07\ m$

**(g)** *Precision of fixation*

$$\mathbf{v} = 10^{-4} \begin{pmatrix} 8\cdot3 \\ 29\cdot6 \\ 0\cdot116 \\ 0\cdot001 \end{pmatrix}$$

$$\sigma_0^2 = (\mathbf{v}^{\mathsf{T}}\mathbf{W}\mathbf{v})/(m - n) = 1\cdot686/(4 - 2) = 0\cdot843$$

$$\sigma_{xx} = \sigma_0^2(\mathbf{A}^{\mathsf{T}}\mathbf{W}\mathbf{A})^{-1} = 10^{-6} \begin{pmatrix} +4\cdot651\ 3 + 0\cdot196\ 8 \\ +0\cdot196\ 8 + 5\cdot408\ 2 \end{pmatrix}$$

$$\sigma_e^2 = 4\cdot651\ 3\ \text{mm}^2, \qquad \sigma_n^2 = 5\cdot408\ 2\ \text{mm}^2, \qquad \sigma_{en} = +0\cdot196\ 8\ \text{mm}^2$$

The pedal curve is illustrated in Fig 4.12.

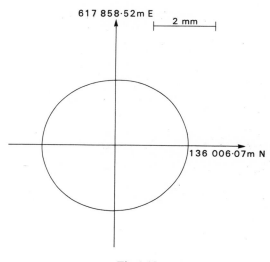

Fig 4.12

## 4.5   The Method of Condition Equations

### 4.5.1   *A Level Net Adjustment by Conditions*
Fig 4.13 represents the level lines 1, 2,..., 10, run between junction points A, B,..., F, to provide the height control for an irrigation scheme survey. The results of the levelling are shown in the table.

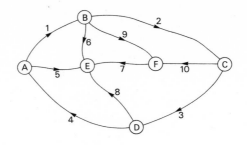

| Line | From | To | Observed height difference (m) | Length of line (km) |
|------|------|-----|-------------------------------|---------------------|
| 1 | A | B | +16·298 | 1·3 |
| 2 | B | C | −17·700 | 1·9 |
| 3 | C | D | − 0·687 | 2·3 |
| 4 | D | A | + 2·086 | 2·7 |
| 5 | A | E | +23·615 | 0·3 |
| 6 | B | E | + 7·304 | 0·9 |
| 7 | F | E | +14·162 | 1·2 |
| 8 | D | E | +25·709 | 1·6 |
| 9 | B | F | − 6·855 | 1·1 |
| 10 | C | F | +10·863 | 0·5 |

Fig 4.13

There are five independent loops in the net, giving five independent conditions to be fulfilled ([22] shows that if there are $S$ stations and $L$ lines measured, then the number of conditions is $L - S + 1$. Here, $L = 10$, $S = 6$ and there are therefore five conditions).

Suppose $X_i$ is the MPV of the difference in height of line $i$. The conditions to be fulfilled can be summarised as in Fig 4.14.

| Circuit | Condition equation |
|---------|-------------------|
| ABEA | $X_1 + X_6 - X_5 = 0$ |
| BFEB | $X_9 + X_7 - X_6 = 0$ |
| BCFB | $X_2 + X_{10} - X_9 = 0$ |
| EFCDE | $-X_7 - X_{10} + X_3 + X_8 = 0$ |
| EDAE | $X_5 - X_8 + X_4 = 0$ |

Fig 4.14

A circuit such as ABFEA gives the condition equation $X_1 + X_9 + X_7 - X_5 = 0$, but this is a linear combination of the first two equations in Fig 4.14 and therefore must not be included. The five condition equations of Fig 4.14 are necessary and sufficient. If the substitution for $X_i$ from (3.20) is made, the condition equations become

$$v_1 - v_5 + v_6 = +0\cdot013$$
$$-v_6 + v_7 + v_9 = -0\cdot003$$
$$v_2 - v_9 + v_{10} = -0\cdot018$$
$$v_3 - v_7 + v_8 - v_{10} = +0\cdot003$$
$$v_4 + v_5 - v_8 = +0\cdot008$$

or $\mathbf{Av} = \mathbf{b}$, where

$$\mathbf{A} = \begin{pmatrix} 1 & 0 & 0 & 0 & -1 & +1 & 0 & 0 & 0 & 0 \\ 0 & 0 & 0 & 0 & 0 & -1 & +1 & 0 & +1 & 0 \\ 0 & +1 & 0 & 0 & 0 & 0 & 0 & 0 & -1 & +1 \\ 0 & 0 & +1 & 0 & 0 & 0 & -1 & +1 & 0 & -1 \\ 0 & 0 & 0 & +1 & +1 & 0 & 0 & -1 & 0 & 0 \end{pmatrix}$$

$$\mathbf{v}^\top = (v_1 \quad v_2 \quad v_3 \quad v_4 \quad v_5 \quad v_6 \quad v_7 \quad v_8 \quad v_9 \quad v_{10})$$

and

$$\mathbf{b}^\top = (+0\cdot013 \quad -0\cdot003 \quad -0\cdot018 \quad +0\cdot003 \quad +0\cdot008)$$

The weight matrix, $\mathbf{W}$, is the diagonal matrix whose typical element, $a_{ii}$, is proportional to the inverse of the length of line $i$ (see section 4.3.1 and equation 4.4).

The correlative normals (3.29) are $(\mathbf{AW}^{-1}\mathbf{A}^\top)\mathbf{k} = \mathbf{b}$, i.e.

$$K^{-1}\begin{pmatrix} +2\cdot5 & -0\cdot9 & 0 & 0 & -0\cdot3 \\ -0\cdot9 & +3\cdot2 & -1\cdot1 & -1\cdot2 & 0 \\ 0 & -1\cdot1 & +3\cdot0 & 0 & 0 \\ 0 & -1\cdot2 & 0 & +5\cdot6 & -1\cdot6 \\ -0\cdot3 & 0 & 0 & -1\cdot6 & +4\cdot6 \end{pmatrix}\begin{pmatrix} k_1 \\ k_2 \\ k_3 \\ k_4 \\ k_5 \end{pmatrix} = \begin{pmatrix} +0\cdot013 \\ -0\cdot003 \\ -0\cdot018 \\ +0\cdot003 \\ +0\cdot008 \end{pmatrix}$$

where $K$ is an arbitrary constant and $\mathbf{k}$ is the $5 \times 1$ matrix of the correlates.

Solution gives

$$\mathbf{k}^\top = (+0\cdot004\,8 \quad -0\cdot001\,6 \quad -0\cdot006\,6 \quad +0\cdot000\,7 \quad +0\cdot001\,8)\,K$$

and substitution for **k** in (3.28) gives the solution **v** as

$$\mathbf{v}^{\mathsf{T}} = (+0{\cdot}006 \quad -0{\cdot}012 \quad +0{\cdot}002 \quad +0{\cdot}007 \quad -0{\cdot}001 \quad +0{\cdot}006$$
$$-0{\cdot}003 \quad -0{\cdot}002 \quad +0{\cdot}006 \quad \pm 0{\cdot}000)$$

so that the adjusted differences in height are as follows

$$X_1 = +16{\cdot}304 \quad X_4 = +2{\cdot}093 \quad X_7 = +14{\cdot}159 \quad X_9 = -6{\cdot}849$$
$$X_2 = -17{\cdot}712 \quad X_5 = +23{\cdot}614 \quad X_8 = +25{\cdot}707 \quad X_{10} = +10{\cdot}863$$
$$X_3 = -0{\cdot}685 \quad X_6 = +7{\cdot}310$$

### 4.5.2 *A Traverse Adjustment by Conditions*

Figure 4.15 illustrates a traverse with $n$ stations between the terminal points A and B. The table explains the symbols used.

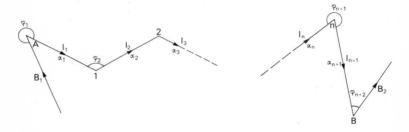

| Symbol | Range | Meaning |
|---|---|---|
| $B_1$ | | Initial, fixed bearing to A |
| $B_2$ | | Final, fixed bearing at B |
| $(E_A, N_A)$ | | Coordinates (fixed) of A |
| $(E_B, N_B)$ | | Coordinates (fixed) of B |
| $\phi_i$ | | Observed horizontal angle |
| $\Phi_i$ | $i = 1, 2,..., n + 2$ | MPV of horizontal angle |
| $\delta\phi_i$ | | 'Correction' to $\phi_i$ to give $\Phi_i$ |
| $l_i$ | | Measured horizontal distance |
| $L_i$ | $i = 1, 2,..., n + 1$ | MPV of horizontal distance |
| $\delta l_i$ | | 'Correction' to $l_i$ to give $L_i$ |
| $\alpha_i$ | | 'Observed' bearing: $\alpha_i = B_1 + \sum\limits_{j=1}^{i} (\phi_j + \pi)$ |
| $A_i$ | $i = 1, 2,..., n + 1$ | MPV of a bearing |
| $\delta\alpha_i$ | | 'Correction' to $\alpha_i$ to give $A_i$ |
| $s_\phi$ | | Standard error of an observed angle |
| $s_l$ | | Standard error of a measured distance |

Fig 4.15

There are three conditions to be fulfilled by the MPVs of the angles and distances: closures in bearings, eastings and northings. The first condition can be expressed as

$$\sum_{i=1}^{n+2} \Phi_i - (n + 2)\pi = B_2 - B_1$$

Substitution of $\Phi_i = (\phi_i + \delta\phi_i)$ and rearrangement gives

$$\sum_{i=1}^{n+2} \delta\phi_i = B_2 - B_1 - \sum_{i=1}^{n+2} (\phi_i - \pi) \qquad (4.31)$$

The second condition can be expressed as

$$\sum_{i=1}^{n+1} L_i \sin A_i = E_B - E_A$$

and substitution of $L_i = l_i + \delta l_i$ and of $A_i = \alpha_i + \delta\alpha_i$ gives

$$\sum_{i=1}^{n+1} (l_i + \delta l_i).\sin (\alpha_i + \delta\alpha_i) = E_B - E_A$$

But

$$(l + \delta l).\sin (\alpha + \delta\alpha) \simeq (l + \delta l)(\sin \alpha + \delta\alpha \cos \alpha)$$
$$\simeq l \sin \alpha + l \,\delta\alpha \cos \alpha + \delta l \sin \alpha$$

Therefore the second condition can be expressed as

$$\sum_{i=1}^{n+1} (l_i \,\delta\alpha_i \cos \alpha_i + \delta l_i \sin \alpha_i) = E_B - E_A - \sum_{i=1}^{n} l_i \sin \alpha_i \quad (4.32)$$

Similarly, the third condition can be expressed as

$$\sum_{i=1}^{n+1} (-l_i \,\delta\alpha_i \sin \alpha_i + \delta l_i \cos \alpha_i) = N_B - N_A - \sum_{i=1}^{n+1} l_i \cos \alpha_i$$

$$(4.33)$$

These three condition equations can be written as

$$\delta\phi_1 + \delta\phi_2 + \cdots + \delta\phi_n + \delta\phi_{n+1} + \delta\phi_{n+2} = b_1$$
$$a_{21} \,\delta\phi_1 + a_{22} \,\delta\phi_2 + \cdots + a_{2\,n+1} \,\delta\phi_{n+1} + \delta l_1 \sin \alpha_1$$
$$+ \,\delta l_2 \sin \alpha_2 + \cdots + \delta l_{n+1} \sin \alpha_{n+1} = b_2$$
$$a_{31} \,\delta\phi_1 + a_{32} \,\delta\phi_2 + \cdots + a_{3\,n+1} \,\delta\phi_{n+1} + \delta l_1 \cos \alpha_1$$
$$+ \,\delta l_2 \cos \alpha_2 + \cdots + \delta l_{n+1} \sin \alpha_{n+1} = b_3$$

or $\mathbf{Av} = \mathbf{b}$ where

$$\mathbf{v} = (\delta\phi_1 \; \delta\phi_2 \; \cdots \; \delta\phi_{n+2} \; \delta l_1 \; \delta l_2 \; \cdots \; \delta l_{n+1})^{\top}$$

$$a_{2j} = \sum_{k=j}^{n+1} l_k \cos\alpha_k \quad \text{and} \quad a_{3j} = -\sum_{k=j}^{n+1} l_k \sin\alpha_k$$

The matrix $\mathbf{A}$ has dimensions $3 \times (2n + 3)$ and the elements of $\mathbf{b}$ are the respective right-hand sides of (4.31), (4.32) and (4.33), with due regard to the units.

The LS solution is obtained as described in section 3.6.2, taking (with due regard to units)

$$\mathbf{W} = \begin{pmatrix} 1/s_\phi^2 & 0 & \cdots & 0 \\ 0 & 1/s_\phi^2 & \cdots & 0 \\ \vdots & \vdots & & \vdots \\ 0 & 0 & \cdots & 1/s_l^2 \end{pmatrix}$$

where the dimensions of $\mathbf{W}$ are $(2n + 3) \times (2n + 3)$.

As a numerical example, let the fixed data be

$$\begin{array}{ll} E_A = 163\ 208\cdot49 \text{ m} & B_1 = 48°\ 27'\ 30'' \\ N_A = 104\ 375\cdot29 \text{ m} & B_2 = 67°\ 48'\ 48'' \\ E_B = 165\ 074\cdot49 \text{ m} & \\ N_B = 105\ 227\cdot47 \text{ m} & \end{array}$$

and the observed horizontal angles and horizontal distances (with significant systematic errors removed by measurement techniques and by a suitable mathematical model) be

$$\begin{array}{ll} \phi_1 = 203°\ 41'\ 28'' & l_1 = 703\cdot28 \text{ m} \\ \phi_2 = 162°\ 37'\ 21'' & l_2 = 473\cdot29 \text{ m} \\ \phi_3 = 193°\ 18'\ 06'' & l_3 = 687\cdot48 \text{ m} \\ \phi_4 = 170°\ 08'\ 49'' & l_4 = 202\cdot31 \text{ m} \\ \phi_5 = 189°\ 35'\ 52'' & \end{array}$$

The notation of Fig 4.15 is used, and here $n = 3$.

The right-hand sides of (4.31), (4.32) and (4.33) are, respectively,

$$B_2 - B_1 - \sum_{i=1}^{5} (\phi_i - \pi) = -18'' = b_1$$

and $\quad E_B - E_A - \sum_{i=1}^{4} (l_i \sin\alpha_i) = +0\cdot232 \text{ m} = b_2$

and $\quad N_B - N_A - \displaystyle\sum_{i=1}^{4} (l_i \cos \alpha_i) = +0\cdot329 \text{ m} = b_3$

Thus the angular misclosure is $+18''$, that in eastings is $-0\cdot232$ m and that in northings is $-0\cdot329$ m. The nett positional misclosure is $(b_2^2 + b_3^2)^{1/2} = 0\cdot403$ m and the proportional positional misclosure is 1 part in 5132.

The condition equations are given as $\mathbf{Av} = \mathbf{b}$, where

$$
\mathbf{A} = \begin{pmatrix}
1 & 1 & 1 & 1 & 1 & 0 & 0 & 0 & 0 \\
851\cdot9 & 636\cdot3 & 363\cdot3 & 106\cdot5 & 0 & 0\cdot9519 & 0\cdot8169 & 0\cdot9277 & 0\cdot8501 \\
-1865\cdot8 & -1196\cdot3 & -809\cdot7 & -172\cdot0 & 0 & 0\cdot3065 & 0\cdot5768 & 0\cdot3734 & 0\cdot5267
\end{pmatrix}
$$

$$
\mathbf{v} = (\delta\phi_1 \; \delta\phi_2 \; \delta\phi_3 \; \delta\phi_4 \; \delta\phi_5 \; \delta l_1 \; \delta l_2 \; \delta l_3 \; \delta l_4)^\mathsf{T}
$$

and $\quad \mathbf{b} = (-8\cdot727 \times 10^{-5} \text{ rad} + 0\cdot232 \text{ m} + 0\cdot329 \text{ m})^\mathsf{T}$

The standard error of an observed angle $(s_\phi)$ has been calculated as $\pm3''$ and the standard error of a measured length $(s_l)$ is $\pm10$ mm. Therefore the weight of an angle measurement is proportional to $s_\phi^{-2} = 4\cdot727 \times 10^9 \text{ rad}^{-2}$ and the weight of a distance measurement is proportional to $s_l^{-2} = 1 \times 10^{-4} \text{ m}^{-2}$. The weight matrix $\mathbf{W}$ (assuming that the observations are uncorrelated) is therefore given by

$$
\mathbf{W} = K.10^9 \begin{vmatrix}
4\cdot727 & 0 & 0 & 0 & 0 & 0 & 0 & 0 & 0 \\
0 & 4\cdot727 & 0 & 0 & 0 & 0 & 0 & 0 & 0 \\
0 & 0 & 4\cdot727 & 0 & 0 & 0 & 0 & 0 & 0 \\
0 & 0 & 0 & 4\cdot727 & 0 & 0 & 0 & 0 & 0 \\
0 & 0 & 0 & 0 & 4\cdot727 & 0 & 0 & 0 & 0 \\
0 & 0 & 0 & 0 & 0 & 1 \times 10^{-5} & 0 & 0 & 0 \\
0 & 0 & 0 & 0 & 0 & 0 & 1 \times 10^{-5} & 0 & 0 \\
0 & 0 & 0 & 0 & 0 & 0 & 0 & 1 \times 10^{-5} & 0 \\
0 & 0 & 0 & 0 & 0 & 0 & 0 & 0 & 1 \times 10^{-5}
\end{vmatrix}
$$

The correlative normal equations (3.29) are $(\mathbf{AW}^{-1}\mathbf{A}^\mathsf{T})\mathbf{k} = \mathbf{b}$ or

$$
K^{-1} \begin{pmatrix}
1\cdot058 \times 10^{-9} & 4\cdot142 \times 10^{-7} & -8\cdot555 \times 10^{-7} \\
4\cdot142 \times 10^{-7} & 5\cdot851 \times 10^{-4} & -4\cdot076 \times 10^{-4} \\
-8\cdot555 \times 10^{-7} & -4\cdot076 \times 10^{-4} & 12\cdot685 \times 10^{-4}
\end{pmatrix} \begin{pmatrix} k_1 \\ k_2 \\ k_3 \end{pmatrix}
$$

$$
= \begin{pmatrix} -8\cdot727 \times 10^{-5} \\ 0\cdot232 \\ 0\cdot329 \end{pmatrix}
$$

Solution gives

$$\mathbf{k} = K \begin{pmatrix} 703 \cdot 61 \\ 723 \cdot 04 \\ 539 \cdot 10 \end{pmatrix}$$

The residuals are given by (3.28) as $\mathbf{v} = \mathbf{W}^{-1}\mathbf{A}^{\mathsf{T}}\mathbf{k}$. In this case,

$$\mathbf{v} = (-13''\cdot 9 \quad -5''\cdot 0 \quad -4''\cdot 5 \quad +2''\cdot 4 \quad +3''\cdot 1 \quad +0 \cdot 085 \text{ m}$$
$$+0 \cdot 090 \text{ m} \quad +0 \cdot 087 \text{ m} \quad +0 \cdot 090 \text{ m})$$

Note that the solution is independent of $K$ and that in order to obtain this LS solution, it is necessary to know only the *ratios* of the weights. Application of the corrections to the measured angles and distances gives the following MPVs:

$$\Phi_1 = 203° \ 41' \ 14''\cdot 1 \qquad L_1 = 703 \cdot 365 \text{ m}$$
$$\Phi_2 = 162° \ 37' \ 16''\cdot 0 \qquad L_2 = 473 \cdot 380 \text{ m}$$
$$\Phi_3 = 193° \ 18' \ 01''\cdot 5 \qquad L_3 = 687 \cdot 567 \text{ m}$$
$$\Phi_4 = 170° \ 08' \ 51''\cdot 4 \qquad L_4 = 202 \cdot 400 \text{ m}$$
$$\Phi_5 = 189° \ 35' \ 55''\cdot 1$$

It can be seen that the conditions are fulfilled by these values, and they can then be used to compute the MPVs of the coordinates of stations 1, 2 and 3 with the following results:

| Station | Easting (m) | Northing (m) |
|---------|-------------|--------------|
| 1 | 163 877·98 | 104 590·94 |
| 2 | 164 264·64 | 104 864·04 |
| 3 | 164 902·44 | 105 120·86 |

The coordinates are rounded off to the nearest 0·01 m; this is the degree of precision to which the fixed coordinates of A and B were quoted.

## 4.6 The Rejection of Observations

If a quantity is measured several times, as far as possible under the same conditions, the measurements constitute a sample of a random variable. If it is assumed that the variable has a specific PD, certain statements can be made about the sample of values obtained and it is possible to develop criteria for the rejection of any measurement which appears to have been influenced by something other than random fluctuations in the conditions of measurement (section 1.5).

The specific PD taken here will be the normal distribution (section 3.3), since survey measurements have generally been found to resemble this distribution rather than any other.

It has been stated (section 3.3) that the chance of a value between $\xi \pm \sigma$ occurring is approximately 68·3%, etc. The probabilities of other values occurring between stated limits are as follows:

| Value between | $\xi \pm \sigma$ | $\xi \pm 2\sigma$ | $\xi \pm 2{\cdot}58\sigma$ | $\xi \pm 3\sigma$ | $\xi \pm 3{\cdot}29\sigma$ | $\xi \pm 4\sigma$ |
|---|---|---|---|---|---|---|
| Probability (%) | 68·3 | 95·4 | 99 | 99·73 | 99·9 | 99·994 |

Thus the chance of obtaining a value outside the range ($\xi \pm 3{\cdot}29\sigma$), for example, is 1 in 1000, indicating good grounds for rejecting such a value. However, in practice, $\xi$ and $\sigma$ are unknown, so that the probabilities tabulated above do not apply. It is therefore necessary to deduce probabilities based on their unbiased estimators $\bar{x}$ and $s$, respectively, which are given by (2.47) and (2.52).

Suppose that a variable $\tau$ is defined by $\tau_i = (x - \bar{x})/s$, where $x_i$ is any one of $n$ normally distributed measured values, $\bar{x}$ is the arithmetic mean of the $n$ measured values and $s$ is defined by $s^2 = [(x_i - \bar{x})^2]/n$, the square brackets denoting summation. The three graphs in Fig 4.16

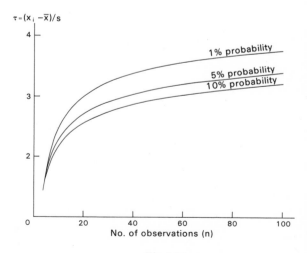

Fig 4.16

give the largest value of $\tau_i$ to be expected from $n$ observations using probabilities of 1%, 5% and 10%. These graphs are based on figures given in [20].

Thus, for example, if ten observations are made and the 1% probability level is chosen as the criterion, then any residual which is greater than 2·48 times $s$ should be rejected. On the other hand, if the 10% probability level is chosen, any residual greater than 2·18 times $s$ should be rejected.

As a numerical example, suppose the micrometer readings in Fig 4.17 to be obtained during a series of rapid 1-second theodolite pointings.

| No | $x_i$ (secs) | $(x_i - \bar{x})$ | $(x_i - \bar{x})^2$ |
|----|------|--------|---------|
| 1 | 41 | −3·2 | 10·24 |
| 2 | 38 | −6·2 | 38·44 |
| 3 | 38 | −6·2 | 38·44 |
| 4 | 58 | +13·8 | 190·44 |
| 5 | 47 | +2·8 | 7·84 |
| 6 | 39 | −5·2 | 27·04 |
| 7 | 46 | +1·8 | 3·24 |
| 8 | 44 | −0·2 | 0·04 |
| 9 | 41 | −3·2 | 10·24 |
| 10 | 50 | +5·8 | 33·64 |

Fig 4.17

$$\bar{x} = 44''\cdot2 \qquad [(x_i - \bar{x})^2] = 359\cdot60 \text{ sec}^2$$

$$s^2 = 35\cdot96 \quad \text{and} \quad s \simeq 6''$$

By reference to Figs 4.16 and 4.17 it can be seen that, if the 1% probability level is the criterion, the maximum residual which can be tolerated is $2\cdot48 \times 6'' \simeq 15''$, and the fourth reading should not be rejected. If the 5% level is the criterion, the maximum residual is $2\cdot28 \times 6'' \simeq 13\cdot7$; the fourth reading is therefore not acceptable, nor is it, of course, acceptable under any other higher probability criterion. The disadvantage of this criterion is that it could fail to locate a reading which has been subject to non-random effects when there is more than one outlying observation [21].

Apart from the foregoing statistical basis for the rejection of observations, it is often possible to have other reasons for rejecting certain measurements, usually arising from a noticeable change in the conditions of measurement in the field. For example, if the sun appears during one observation to an opaque signal, phasing could mean that that particular reading is not subject to the same random fluctuations in the conditions as the other readings; it should therefore be rejected. It is always preferable to have more than one reason for rejecting any one measurement, and remarks made in the field-book at the time of the observations are useful in this respect.

# CHAPTER FIVE

# Coordinate Systems

Since surveying is concerned primarily with measurements to determine the relative positions of points on, above and below the earth's
surface, the coordinate system used to define the positions is of great
importance and must be chosen to provide an adequate resemblance
to the physical reality. In general, the simpler the system, the more
limited it is in its application.

## 5.1 The Spheroidal Coordinate System

The irregular nature of the earth's *topographical surface* prevents it
being used as a reference system for defining positions of points near
it. This surface is almost an oblate spheroid with semi-axes of about
6378 km (equatorial) and 6357 km (polar). (An oblate spheroid is
generated by rotating an ellipse about its minor axis.)

A surface which more closely resembles an oblate spheroid is the
*geoid*. This can be imagined as being the surface which would be taken
up by the mean level of the sea if it were allowed to pass through the
earth's crust in narrow frictionless channels. It can, however, be better
defined as that equipotential surface arising from both the earth's
attraction and rotation which coincides with the mean level of the sea
in deep water. The geoid has one useful property: at any point near it,
a plumb-bob string indicates the direction of the geoidal normal, or
*vertical*.

Departures of the geoid from a spheroid are small (of the order of 10 m linearly and 10″ in inclination). The measurement of these deviations is part of the subject of geodesy and their interpretation is part of geophysics. As far as this book is concerned, it is assumed that the geoid and the *spheroid of reference* coincide throughout any given region. Fig 5.1 illustrates these three surfaces.

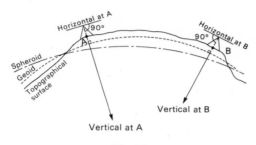

Fig 5.1

For any given point on the topographical surface there will be a corresponding point on the spheroid at the intersection of the vertical at the point (which is here assumed normal to the spheroid as well as to the geoid) and the spheroidal surface.

Fig 5.2 illustrates a topographical point (P′) with its projection onto the spheroid at P.

NS is the *minor axis* of the spheroid, ON = OS = b.
EQ is the *major axis*, OE = OQ = a.
EABQ is the *equator* and its plane is the *equatorial plane.*
PM is the *spheroidal normal* (or *vertical*, if the geoid and spheroid coincide) at P.
NPAS is the *meridian* through P.
NGBS is a *reference meridian* through a datum point such as G.
UPT is the *parallel of latitude* through P; it is the intersection of a plane through P, parallel to the equatorial plane, with the spheroidal surface.

The *spheroidal latitude* ($\phi$) of P is defined as the angle between the spheroidal normal at P and a plane parallel to the equatorial plane, i.e. $\phi = 90° - \angle NMP$.

The *spheroidal longitude* ($\lambda$) of P is defined as the angle between the plane of the meridian through P and the plane of the meridian through the reference point G, i.e. $\lambda = \angle AOB$.

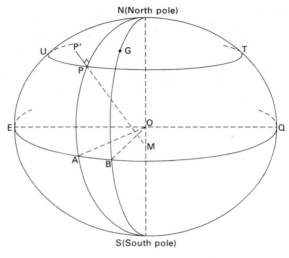

Fig 5.2

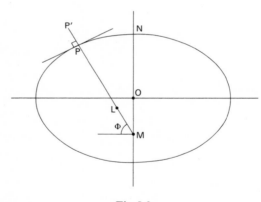

Fig 5.3

Fig 5.3 shows a meridional section containing a given point, P, having latitude $\phi$.

The radius of curvature in the meridian at P is $PL(=\rho)$ and

$$\rho = \frac{a(1 - e^2)}{(1 - e^2 \sin^2 \phi)^{3/2}} \qquad (5.1)$$

The radius of curvature at right angles to the meridian is PM ($= v$) and

$$v = \frac{a}{(1 - e^2 \sin^2 \phi)^{1/2}} \tag{5.2}$$

The radius of curvature $R$ at any latitude $\phi$ and in any azimuth $\alpha$ is given by Euler's theorem:

$$\frac{1}{R_\alpha} = \frac{\cos^2 \alpha}{\rho} + \frac{\sin^2 \alpha}{v} \tag{5.3}$$

The mean radius of curvature at any point is given by $(\rho v)^{1/2}$. This expression and (5.1), (5.2) and (5.3) are derived, for example, in [1]. These quantities appear in several expressions for spheroidal computation and tabulated values are published for various spheroids.

There have been many determinations of the size and shape of the spheroid which most closely resembles the geoid in specific areas. A reference spheroid is usually defined in size and shape by the specification of two of the following parameters of the generating ellipse:

the major axis, $2a$
the minor axis, $2b$
the flattening, $f = (a - b)/a$
the eccentricity, $e = (a^2 - b^2)/a^2$

The results of some determinations are given in Fig 5.4.

| Name | Date | $a$ (metres) | $1/f$ | Remarks |
|---|---|---|---|---|
| Eratosthenes | 250 BC | $6\cdot3 \times 10^6$ approx. | $\infty$ | First known determination |
| Cassini | AD 1718 | 6 361 000 approx. | $-95$ | Prolate spheroid |
| Everest | 1830 | 6 377 276 | 300·80 | India, Burma |
| Bessel | 1841 | 6 377 397 | 299·15 | Germany, Russia |
| Airy | 1849 | 6 377 491 | 299·32 | Great Britain |
| Clarke | 1866 | 6 378 206 | 294·98 | North America |
| ,, | 1880 | 6 378 301 | 293·47 | South, East and West Africa |
| Hayford | 1910 | 6 378 388 | 297·00 | 'International' |
| Reference Ellipsoid } | 1967 | 6 378 160 | 298·247 | Recommended by International Association of Geodesy |

Fig 5.4

When distances and angles measured between points on the topographic surface are to be used to compute positions on the reference spheroid, certain 'corrections' have to be applied to the measured values before they are used in spheroidal computation. Such 'corrections' are considered in Chapter 6. The formulae for computation of position on the reference spheroid are not simple.

The choice of formula will depend upon

(a) the accuracy required
(b) the latitude of the survey
(c) the lengths of the lines
(d) the computing facility available.

The national survey organisations of most countries publish tables of spheroidal constants to assist in such computations, and generally the national primary framework is computed and adjusted by least squares on the spheroid of reference used by that particular country.

For an introduction to spheroidal computation, the reader can refer to [1]. Details of various formulae, their derivations and applications are given in [11].

## 5.2 Rectangular Cartesian Coordinates

The (XYZ) system is defined within the region of a specific project. Since the direction of gravity can be readily determined (by means of a plumb-bob string or a spirit level) it is convenient to define one axis (customarily OZ) as parallel to the (upward) direction of gravity in the area. The XY plane is thus horizontal. The OY axis is customarily taken to be directed towards north. It is convenient to define the origin so that all points in the region have positive coordinates. Thus, a typical point in the region such as P in Fig 5.5 will have coordinates $(X, Y, Z)$ or $(E, N, h)$, referred to a local origin and a local 'north'.

Such a coordinate system is satisfactory only when the variation in the direction of gravity over the area under consideration is small enough to be ignored.

It has been customary to think in terms of eastings and northings on the one hand, and of heights on the other, rather than in terms of $X$, $Y$ and $Z$ spatial coordinates. The reason for this is that survey instrumentation and the computing methods generally available have made this by far the simplest method. Modern developments in both instrumentation and computing may call for a reappraisal of this convention and it is possible that, in future, the use of $(X, Y, Z)$ for surveying, design and setting out will be more widespread.

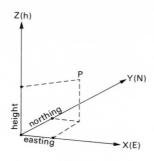

Fig 5.5

The conversion from one plane rectangular system to another arises, for example, when a survey computed on a local coordinate system is to be converted to another plane rectangular system (for example, the National Grid). It also arises when the coordinates of points to be set out are referred to axes defined in terms of the construction itself (e.g. a dam axis often serves as a reference for design parameters). Then, a transformation of the design coordinates to the ground survey coordinate system is often necessary prior to setting out.

In Fig 5.6, P is a point with coordinates $(x, y)$ in the secondary $(oxy)$ system. The coordinates of P $(X, Y)$ in the primary (OXY) system are required. The secondary system can be made coincident with the primary system by a clockwise rotation, $\theta$, and a translation $(-X_0, -Y_0)$ and a magnification $\lambda$ in each direction.

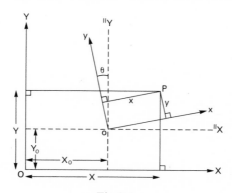

Fig 5.6

Then: $$X - X_0 = \lambda(x \cos \theta - y \sin \theta) \qquad (5.4)$$

and $$Y - Y_0 = \lambda(x \sin \theta + y \cos \theta) \qquad (5.5)$$

Or $$\begin{pmatrix} X \\ Y \end{pmatrix} = \begin{pmatrix} X_0 \\ Y_0 \end{pmatrix} + \begin{pmatrix} a & -b \\ b & a \end{pmatrix} \begin{pmatrix} x \\ y \end{pmatrix} \qquad (5.6)$$

where $a = \lambda \cos \theta$ and $b = \lambda \sin \theta$.

Equation (5.6) expresses the relation between the coordinates $(X, Y)$ of a point in one system and those $(x, y)$ of the same point in another system. For a given transformation there are four constants $(X_0, Y_0, a$ and $b)$; if the coordinates of two different points P and Q are known in both systems, this information gives four equations for the unique determination of the four unknowns.

Suppose P and Q are the two points with coordinates $(X_P Y_P)$ and $(X_Q Y_Q)$ in the primary system and $(x_P y_P)$ and $(x_Q y_Q)$ in the secondary system. Then

$$X_Q = X_0 + ax_Q - by_Q \qquad (5.7)$$
$$Y_Q = Y_0 + bx_Q + ay_Q \qquad (5.8)$$
$$X_P = X_0 + ax_P - by_P \qquad (5.9)$$
$$Y_P = Y_0 + bx_P + ay_P \qquad (5.10)$$

(5.9) − (5.7) gives

$$(X_P - X_Q) = (x_P - x_Q)a - (y_P - y_Q)b \qquad (5.11)$$

and (5.10) − (5.8) gives

$$(Y_P - Y_Q) = (x_P - x_Q)b + (y_P - y_Q)a \qquad (5.12)$$

Therefore $$\begin{pmatrix} X_P - X_Q \\ Y_P - Y_Q \end{pmatrix} = \begin{pmatrix} x_P - x_Q & -y_P + y_Q \\ y_P - y_Q & x_P - x_Q \end{pmatrix} \begin{pmatrix} a \\ b \end{pmatrix}$$

and $$\begin{pmatrix} a \\ b \end{pmatrix} = \begin{pmatrix} x_P - x_Q & -y_P + y_Q \\ y_P - y_Q & x_P - x_Q \end{pmatrix}^{-1} \begin{pmatrix} X_P - X_Q \\ Y_P - Y_Q \end{pmatrix}$$

$$= \frac{1}{d_{PQ}^2} \begin{pmatrix} x_P - x_Q & y_P - y_Q \\ -y_P + y_Q & x_P - x_Q \end{pmatrix} \begin{pmatrix} X_P - X_Q \\ Y_P - Y_Q \end{pmatrix}$$

where $$d_{PQ}^2 = (x_P - x_Q)^2 + (y_P - y_Q)^2 \neq 0,$$

since P and Q are different points.

Thus

$$a = \{(x_P - x_Q)(X_P - X_Q) + (y_P - y_Q)(Y_P - Y_Q)\}/d_{PQ}^2 \tag{5.13}$$

and

$$b = \{-(y_P - y_Q)(X_P - X_Q) + (x_P - x_Q)(Y_P - Y_Q)\}/d_{PQ}^2 \tag{5.14}$$

Substitution of these values in (5.7) and (5.8) enables $X_0$ and $Y_0$ to be determined with an independent check from (5.9) and (5.10).

If there are $n$ points $(n > 2)$ with known coordinates in both systems, then the principle of least squares (see Chapter 3) can be used to determine the most probable values (MPVs) of the transformation constants $X_0$, $Y_0$, $a$ and $b$. This is best done by translating the two axis systems so that their (common) origin is at the centroid (G) of the $n$ points whose coordinates in both systems are known. If this centroid has coordinates $(X_G, Y_G)$ and $(x_G, y_G)$ in the original systems then

$$X_G = [X_i]/n, \qquad\qquad Y_G = [Y_i]/n$$

and

$$x_G = [x_i]/n, \qquad\qquad y_G = [y_i]/n$$

where the summations are for $i = 1$ to $n$. After translation of the two sets of axes so that their common origin is at G, the coordinates of any point $P_i$ are then given by $\bar{X}_i = X_i - X_G$, $\bar{Y}_i = Y_i - Y_G$, $\bar{x}_i = x_i - x_G$ and $\bar{y}_i = y_i - y_G$. The transformation (5.6) is now

$$\begin{pmatrix} \bar{X} \\ \bar{Y} \end{pmatrix} = \begin{pmatrix} a & -b \\ b & a \end{pmatrix} \begin{pmatrix} \bar{x} \\ \bar{y} \end{pmatrix}$$

If $a$ and $b$ are regarded as unknowns, and the coordinates $\bar{X}_i$, $\bar{Y}_i$ as measurements with residuals $v_{X_i}$ and $v_{Y_i}$ respectively, then there are $2n$ observation equations:

$$\begin{pmatrix} \bar{x}_1 & -\bar{y}_1 \\ \vdots & \vdots \\ \bar{x}_n & -\bar{y}_n \\ \bar{y}_1 & \bar{x}_1 \\ \vdots & \vdots \\ \bar{y}_n & \bar{x}_n \end{pmatrix} \begin{pmatrix} a \\ b \end{pmatrix} = \begin{pmatrix} \bar{X}_1 \\ \vdots \\ \bar{X}_n \\ \bar{Y}_1 \\ \vdots \\ \bar{Y}_n \end{pmatrix} + \begin{pmatrix} v_{X_1} \\ \vdots \\ v_{X_n} \\ v_{Y_1} \\ \vdots \\ v_{Y_n} \end{pmatrix}$$

or $\mathbf{Ax} = \mathbf{k} + \mathbf{v}$. From section 3.6.1, the LS solution (taking $\mathbf{W} = \mathbf{I}$)

is $\mathbf{x} = (\mathbf{A}^\top \mathbf{A})^{-1} \mathbf{A}^\top \mathbf{k}$, i.e.

$$\begin{pmatrix} a \\ b \end{pmatrix} = \begin{pmatrix} [\bar{x}_i^2 + \bar{y}_i^2] & 0 \\ 0 & [\bar{x}_i^2 + \bar{y}_i^2] \end{pmatrix}^{-1} \begin{pmatrix} [\bar{x}_i \bar{X}_i + \bar{y}_i \bar{Y}_i] \\ [-\bar{y}_i \bar{X}_i + \bar{x}_i \bar{Y}_i] \end{pmatrix}$$

so that $\qquad\qquad a = [\bar{x}_i \bar{X}_i + \bar{y}_i \bar{Y}_i]/[\bar{x}_i^2 + \bar{y}_i^2]$

and $\qquad\qquad b = [\bar{x}_i \bar{X}_i - \bar{y}_i \bar{X}_i]/[\bar{x}_i^2 + \bar{y}_i^2]$

Finally, by translation of the axes to their original origins, substitution in (5.6) gives

$$X_0 = X_G - ax_G + by_G$$

and $\qquad\qquad Y_0 = Y_G - bx_G - ay_G$

## 5.3 The Orthomorphic Projection

The formulae for computation on a spheroidal surface are not simple and it is generally more convenient to compute on a plane rectangular coordinate system. Also, since maps are produced on a plane surface, it is necessary to define a projection of the spheroidal surface onto a plane in a specific area. If any point in the region of a survey has spheroidal coordinates $(\phi, \lambda)$, the projection is defined by

$$E = f_1(\phi, \lambda) \quad \text{and} \quad N = f_2(\phi, \lambda) \qquad (5.15)$$

where $f_1$ and $f_2$ are specified functions and $E$ and $N$ are the coordinates of the same point in the plane rectangular coordinate system of the projection.

The functions $f_1$ and $f_2$ are chosen so that some property is preserved after projection; for example, areas can remain the same, or bearings from a given point can be maintained. Since surveying involves the measurement of distances and angles on an approximately spheroidal surface, the projection chosen for computation should distort these measured quantities as little as possible. Such a projection is called *orthomorphic*. In order to illustrate the properties of an orthomorphic projection, a sphere (radius $R$) will be considered instead of the spheroid as this leads to simpler expressions.

Fig 5.7 shows a short line PQ on the (spherical) earth, with components $\delta\phi$ and $\delta\lambda$. The lengths of the arcs PS and QS are, respectively, $R \cos \phi . \delta\lambda$ and $R . \delta\phi$. Suppose that, after projection according to (5.15), these elementary lengths become $\delta E$ and $\delta N$ respectively.

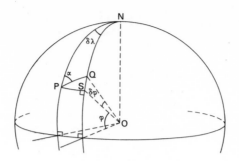

Fig 5.7

The *scale factor* in the EW direction is defined as

$$K_E = \delta E / R \cos \phi . \delta \lambda \qquad (5.16)$$

and that in the NS direction is defined as

$$K_N = \delta N / R . \delta \phi \qquad (5.17)$$

The projection will be orthomorphic if $K_E = K_N$. In such a case,

$$\frac{\delta E}{R \cos \phi . \delta \lambda} = \frac{\delta N}{R . \delta \phi}$$

Therefore

$$\frac{\delta E}{\delta N} = \frac{\delta \lambda \cos \phi}{\delta \phi} \qquad (5.18)$$

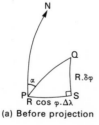

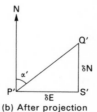

Fig 5.8

An interpretation of (5.18) is seen from Fig 5.8 to be that $\tan \alpha = \tan \alpha'$, so that $\alpha = \alpha'$. Thus the angles between points separated by elementary distances are preserved, and the shapes of small areas remain the same after projection. The projection is therefore called *orthomorphic* ( ὀρθόσ μορφή = right shape).

It can be seen from the foregoing that an orthomorphic projection is particularly suited to survey computations since, over small distances,

(a) $K_F = K_N$, and spheroidal distances from a given point to other nearby points bear a constant ratio to the corresponding distances between the points after projection; and

(b) angles measured between points on the spheroid are the same as the angles between the corresponding points after projection.

Examples of the magnitudes of the discrepancies over longer lines are given in sections 5.4.1 and 5.4.2.

## 5.4 The Transverse Mercator (TM) Projection

This is an orthomorphic projection of the spheroid which in one form or another is in widespread use. The TM projection of a sphere is considered here for simplicity. The TM projection of a spheroid is described fully in [18].

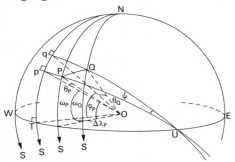

Fig 5.9

In Fig 5.9, a point P is shown with latitude $\phi_P$ and longitude $\Delta\lambda_P$ referred to a specific meridian NqpTS the *central meridian* (CM). Suppose the secondary to the CM which passes through P cuts the CM at $p$. The pole of the CM is at U, on the equator WE.

Suppose that plane rectangular coordinates $(e_P, n_P)$ of P are defined by

$$e_P = \text{arc } pP = R\theta_P \qquad (5.19)$$

and

$$n_P = \text{arc } Tp = R\omega_P \qquad (5.20)$$

where $R$ is the radius of the sphere. For another point, Q, the plane rectangular coordinates will be similar:

$$e_Q = R\theta_Q \quad \text{and} \quad n_Q = R\omega_Q \tag{5.21}$$

In order to see whether (5.19) and (5.20) lead to an orthomorphic projection, consider Fig 5.10 where P$n$ is part of the small circle through P which is parallel to the CM. The distance PQ on the sphere has components PV and VQ in the projection $n$ and $e$ directions respectively.

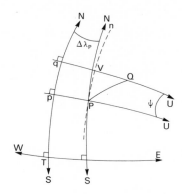

Fig 5.10

Therefore the scale factors are

$$K_E = \frac{\text{arc } qQ - \text{arc } pP}{\text{arc } VQ} = \frac{R\theta_Q - R\theta_P}{R(\theta_Q - \theta_P)} = 1$$

and

$$K_N = \frac{\text{arc } pq}{\text{arc } PV} = \frac{R(w_Q - w_P)}{\psi R \cos \theta_P} = \frac{R\psi}{R\psi \cos \theta_P} = \sec \theta_P$$

so that this projection (Cassini's) is not orthomorphic. It can be made orthomorphic if $K_E$ is put equal to $\sec \theta_P$. Thus for elementary components $\delta\theta$ and $\delta e$:

$$K_E = \frac{\delta e}{R.\delta\theta} = \sec \theta_P$$

so that

$$e_P = \int_0^{\theta_P} R \sec \theta \, d\theta = R \log_e \tan (\pi/4 + \theta_P/2) \tag{5.22}$$

Equations (5.20) and (5.22) enable the orthomorphic projection coordinates $(e_P, n_P)$ to be computed, given the coordinates $(\omega_P, \theta_P)$; these latter are themselves functions of the latitude and difference in longitude $(\phi_P, \Delta\lambda_P)$. Projection tables are published to assist in the conversion of the spheroidal coordinates $(\phi, \Delta\lambda)$ to projection co-ordinates and vice versa. Numerical examples can be found in [1] and [4].

With the increasing use of computer programmes for conversion and other computations, published tables are of less importance. Formulae to provide a basis for a programmed conversion are given in or referred to in [11].

In Fig 5.10, the angle at P between PN and P$n$ is the *convergence* $(c_P)$ and is the angle between the spherical meridian at P and the direction of grid north. This is shown, in [1] for example, to be given by

$$\tan c_P = \tan \Delta\lambda_P \sin \phi_P \qquad (5.23)$$

in the case of a sphere.

### 5.4.1 *The Scale Factor*

It has been shown that the TM projection is orthomorphic and that the scale factor is sec $\theta$ (section 5.4). This means that, when a distance is measured and reduced to the spheroid, it must be multiplied by sec $\theta$ before it is used in plane computation on the projection.

The scale distortion becomes considerable at even a $3°$ angular distance from the CM: then, sec $\theta \simeq 1\cdot0014$ so that a distance $d$ on the spheroid becomes $1\cdot0014d$ on the projection, thereby having a distortion of about 1 part in 700. This is significant even for scaling from a plan. This 'error' can be reduced by limiting the EW extent of a projection to an acceptable level and also by making the scale factor along the CM $(K_0)$ equal to, say, $0\cdot9996$ instead of unity. In the latter case, the distribution of the scale factor across the EW extent of the projection is illustrated in Fig 5.11.

Strictly, the scale factor applies *at a point*, but for lines of limited length it may be assumed that the scale factor at the centre of the line can be applied to the whole of the line without introducing significant error. Since the scale factor $K$ is given by

$$K = \sec \theta \qquad (5.24)$$

an error $\delta K$ in $K$ arising from an error in $\theta$ is given by

$$\delta K = \tan \theta \sec \theta \, \delta\theta$$

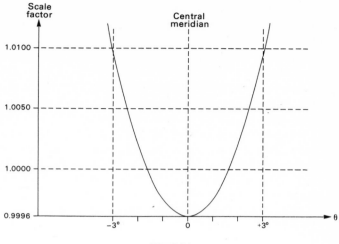

Fig 5.11

Therefore $\delta K \simeq (\theta + \theta^3/3 + \cdots)(1 + \theta^2/2 + \cdots)^{-1}\,\delta\theta$

$$\simeq \theta\,.\,\delta\theta \qquad (5.25)$$

since $\theta \leqslant 3°$, say.

Suppose a spheroidal distance $2L$ is measured, running roughly EW near $\theta = 3°$. The projection distance, $L'$, is given by $L' = 2LK$ and if $K$ has an error $\delta K$ the error in $L'$ will be

$$\delta L' \simeq 2L\,.\,\theta\,.\,\delta\theta \qquad (5.26)$$

from (5.25). Since the line runs roughly EW, $\delta\theta$ is approximately equal to $L/R$ so that (5.26) becomes

$$\delta L' \simeq 2L^2\theta/R$$

If the maximum error in $L$ which can be tolerated is 10 mm, then

$$10 \simeq \frac{2L^2\theta}{R} \times 10^6 \qquad (5.27)$$

where $L$ and $R$ are in km and $\theta = 3\pi/180$ radians. If $R$ is taken as 6370 km, $2L$ is found from (5.27) to be 1·6 km. Thus the scale factor at the centre of a line 1·6 km long can be taken over the whole line without introducing an error greater than 10 mm (1 part in 160 000). (Note that neglect of the scale factor altogether would cause an error in the projection distance of 1 part in 700.)

When the length of a line AB is such that use of the scale factor at only its mid-point (C) is not sufficiently accurate, the scale factor can be computed from

$$K = \tfrac{1}{6}(K_A + 4K_C + K_B) \tag{5.28}$$

where $K_A$, $K_B$ and $K_C$ are the scale factors at A, B and C respectively.

### 5.4.2 *Angular Distortion*

Orthomorphism means that angles on the projection are the same as the corresponding angles on the spheroid, but this applies in theory only to angles between elementary side lengths. When long lines are observed, it is sometimes necessary to determine a 'correction' to the observed angles before using them for plane computation on the TM projection.

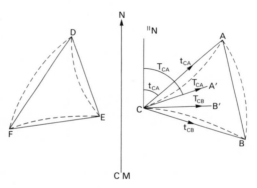

Fig 5.12

In Fig 5.12, two triangles on the projection (ABC and DEF) are shown, one on each side of the CM. The grid bearing of line CA, for example, is $t_{CA}$. The shortest line on the spheroid (the *geodesic*) between C and A, however, plots as the pecked line CA. In the absence of any atmospheric effects, the theodolite at C, when sighted towards A, points in the direction of the tangent to the pecked line at C (CA'), which has a grid bearing $T_{CA}$.

The difference between the two directions, $(t - T)_{CA}$, is known as the *arc to chord* 'correction' or the $(t - T)$ 'correction'. There is a similar correction $(t - T)_{CB}$ for the direction CB, so that the observed spheroidal angle (A'ĈB') becomes the plane angle (AĈB):

$$A\hat{C}B = A'\hat{C}B' + (t - T)_{CA} + (t - T)_{CB} \tag{5.29}$$

where the sign of any $(t - T)$ can be positive or negative. A formula [11] suitable for most calculations of $(t - T)_{CA}$, for example, is

$$(t - T)_{CA} = -\frac{(2E_C + E_A)(N_A - N_C)}{6\rho v K_0^2} \text{ radians} \qquad (5.30)$$

The geodesic always lies on that side of the line on the projection where the scale is greater, so that the geodesics are always concave to the CM. The sign of the $(t - T)$ correction is best determined by inspection of a diagram such as Fig 5.12.

From (5.30) it can be seen that, in general, $(t - T)_{CA} \neq (t - T)_{AC}$.

In order to get an idea of the likely maximum magnitude of a $(t - T)$ 'correction', consider a line near the edge of the 3° belt of projection where $E_A = E_C \simeq 333$ km, from (5.22), taking $R = 6360$ km. Suppose the maximum error acceptable in a plane direction is $1''$, then (5.30) becomes (without the sign)

$$\frac{1}{2 \cdot 06 \times 10^5} \simeq \frac{(999)(N_A - N_C)}{6R^2 \times 1}$$

and $(N_A - N_C) \simeq 1 \cdot 2$ km. Thus, in general, lines of less than 1 km introduce $(t - T)$ 'corrections' of less than $1''$.

For a given triangle, the sum of the $(t - T)$ corrections is equal to the *spherical excess* (see section 6.6.2).

### 5.4.3  Convergence

Convergence has been defined in section 5.4, Fig 5.10 and (5.23). When astronomical or sun azimuths are computed from timed theodolite observations, the azimuth is reckoned from true north (TN). If such a measurement is to be used for computing on the projection,

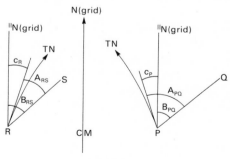

Fig 5.13

the true north azimuth ($A$) must be converted to grid bearing ($B$) by the application of convergence ($c$) (Fig 5.13). The convergence of the meridians towards the CM means that for a line E of the CM, such as PQ in Fig 5.13,

$$B_{PQ} = A_{PQ} - c_P \qquad (5.31)$$

and for a line W of the CM, such as RS in Fig 5.13,

$$B_{RS} = A_{RS} + c_Q$$

From (5.23) it can be seen that

$$c_P'' \simeq \Delta\lambda_P'' \sin \phi_P$$

so that the convergence can be as large as $1°$ or more and it is generally necessary to apply it to observed azimuths.

### 5.4.4 *The Universal Transverse Mercator Projection (UTM)*
Owing to the large scale error of the TM projection away from the CM, the world is divided into sixty zones, each $6°$ wide in longitude, with central meridians at $177°$W for zone 1, $171°$W for zone 2, etc. The scale along each CM is $0·9996$ and, within each zone, the grid easting for the CM is 500 000 m, so as to avoid negative coordinates.

Tables for conversion from grid to spheroidal coordinates and vice versa, and for calculation of scale factor, ($t - T$) and convergence values, are produced for various spheroids—see, for example, [6] and [7]. Formulae which provide a basis for programmed computing can be found, for example, in [11].

### 5.4.5 *The National Grid*
The National Grid is the name of the TM projection used in Great Britain. The CM is at $2°$W and the scale factor along it is $0·9996$. Instead of measuring northings from the equator, as in the UTM, the origin is $49°$N. In order to avoid negative eastings, the grid origin is 400 km W of the CM, and, in order not to exceed six digits in the metres of northings, the grid origin is 100 km N of $49°$N.

Tables based on the TM projection of the Airy 1849 spheroid are published [5], to assist in computation. Formulae which provide a basis for programmed computing can be found or are referred to in [11].

# CHAPTER SIX

# Mathematical Models for Basic Surveying Measurements

Before measurements are analysed according to the statistical principles which are outlined in Chapter 2 and illustrated in Chapters 3 and 4, a mathematical model of the measuring process must be devised. This model should bear an adequate resemblance to the physical reality and depend also upon the coordinate system used to define positions.

In this chapter, certain basic surveying measurements are considered and corresponding mathematical formulae are given which, when included in the mathematical model, reduce systematic errors. It is important in any particular case to include only those terms which are necessary and sufficient for the removal of significant systematic error: unnecessary terms are wasteful of computing time and insufficient terms lead to a departure of the model from reality, with the consequent systematic error.

These terms are often called 'corrections', and this expression is used here; however, the concomitant belief that, once they are applied, the results are 'correct', is to be avoided.

## 6.1 Linear Measurement (General)

There are three methods of distance measurement: mechanical, electromagnetic and optical. In each of these methods, readings are obtained which enable the distance to be computed. Sections 6.2, 6.3

and 6.4 deal with the ways in which these readings are transformed to give a value for $D$, the rectilinear distance between the terminals of a line.

Once a value for $D$ has been obtained, it may be (and generally is) necessary to modify the value to take account of the coordinate system used for defining positions.

Sub-sections 6.1.1, 6.1.2 and 6.1.3 describe modifications to the rectilinear distance $D$ (whether $D$ is obtained by mechanical, optical or electromagnetic methods) which may be needed for an adequate mathematical model.

### 6.1.1  Slope Correction Term

It is generally necessary to consider this term, no matter which co-ordinate system is to be used.

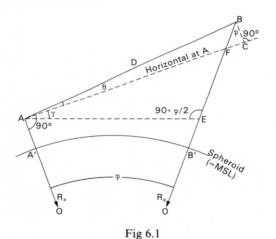

Fig 6.1

In Fig 6.1, AB is the known rectilinear distance $D$, and the vertical angle at A, corrected for refraction (section 6.5.1), is $\theta$. The angle subtended at O, the centre of curvature of MSL, is $\phi$. From triangle ABE it can be seen that

$$AE = D \cos (\theta + \phi) \sec (\phi/2) = c_A$$

so that the correction for slope to give the chord distance $c_A$ at height $h_A$ is $C_S = c_A - D$. Thus,

$$C_S = -D\{1 - \cos (\theta + \phi) \sec (\phi/2)\} \tag{6.1}$$

Generally, $\phi$ can be put equal to zero, but, to cite an extreme case, if $D = 30$ km and $\theta = +3°$ (corresponding to a height difference of about 1600 m between A and B), the effect of putting $\phi = 0$ means that $C_S$ is wrong by about 8 m (i.e. 1 part in 3750).

A value for $\phi$ can be found from $\phi \simeq c_A/(R_\alpha + h_A)$ where $D$ can be used for $c_A$, $R_\alpha$ is the radius of curvature (5.3) and $h_A$ is the height of A above the spheroid. Subsequently, a revised value for $c_A$ can be used. For most purposes, however,

$$C_S \simeq -D(1 - \cos \theta) = -D \text{ versine } \theta \qquad (6.2a)$$

Sometimes the heights of the terminals of the line are known, rather than the vertical angle. In such a case (Fig 6.2) the known distance $D \, (= AB)$ can be reduced to the chord distance $c_A$ using the heights of A and B ($h_A = AA'$ and $h_B = BB'$ respectively).

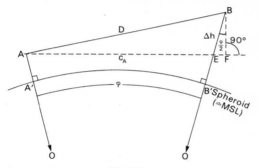

Fig 6.2

In Fig 6.2, $BE = (h_B - h_A) = \Delta h$ and $BF = \Delta h \cos (\phi/2)$. Therefore,

$$
\begin{aligned}
c_A &= \{D^2 - \Delta h^2 \cos^2 (\phi/2)\}^{1/2} - \Delta h \sin (\phi/2) \\
&= D\{1 - (\Delta h^2/D^2) \cos^2 (\phi/2)\}^{1/2} - \Delta h \sin (\phi/2) \\
&\simeq D\{1 - (\Delta h^2/2D^2) \cos^2 (\phi/2) - (\Delta h^4/8D^4) \cos^4 (\phi/2)\} \\
&\qquad\qquad\qquad\qquad\qquad\qquad\qquad - \Delta h \sin (\phi/2)
\end{aligned}
$$

and the correction for slope, $C_S = c_A - D$, is

$$C_S \simeq -(\Delta h^2/2D) \cos^2 (\phi/2) - (\Delta h^4/8D^3) \cos^4 (\phi/2) - \Delta h \sin (\phi/2)$$

As before, $\phi$ can generally be put equal to zero so that

$$C_S \simeq -(\Delta h^2/2D) - (\Delta h^4/8D^3) \qquad (6.2b)$$

with the term in $\Delta h^4$ often insignificant. Thus, the horizontal distance, $c_A$, at height $h_A$ is given by

$$c_A = D + C_S \tag{6.3}$$

where $C_S$ is from either (6.2a) or (6.2b). As an approximate formula, (6.2b) should be more accurate than (6.2a), since the curvature of the earth should have been taken into account in evaluating the term $\Delta h$ in (6.2b), whereas it is ignored entirely in (6.2a).

### 6.1.2  *Reduction to the Spheroid*
If a distance is to be used for spheroidal computation or computation on a projection such as the Transverse Mercator, then the value from (6.3) must be reduced to the spheroid.

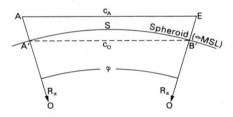

Fig 6.3

In Fig 6.3, the notation is the same as for Fig 6.2. The spheroidal distance is the arc A′B′ ($= S$), which is assumed to be circular with radius $R_\alpha$ given by (5.3). AE $= c_A$ is given by (6.3) and is reduced first to the MSL chord distance A′B′ $= c_0$ by a *MSL term*:

$$c_0/c_A = R_\alpha/(R_\alpha + h_A)$$

Therefore,     $c_0 = c_A\{1 + (h_A/R_\alpha)\}^{-1}$

$$= c_A\{1 - (h_A/R_\alpha) + (h_A/R_\alpha)^2 - \cdots\}$$

and the MSL term, $C_M$ to be applied to $c_A$ to give $c_0$ is

$$C_M = c_0 - c_A \simeq -c_A(h_A/R_\alpha) \tag{6.4}$$

Reduction of this MSL chord length to the spheroid is carried out by the inclusion of a further term $C_R$, given by $C_R = S - c_0$ where $S$ is the arc A′B′.

$$S = R_\alpha\phi \quad \text{and} \quad c_0 = 2R_\alpha \sin(\phi/2)$$

Therefore,
$$c_0 = 2R_\alpha \sin (S/2R_\alpha)$$
$$= 2R_\alpha\{(S/2R_\alpha) - (S^3/48R_\alpha^3) + \cdots\}$$
$$\simeq S - (S^3/24R_\alpha^2)$$

Therefore,
$$C_R \simeq S^3/24R_\alpha^2 \simeq c_0^3/24R_\alpha^2 \tag{6.5}$$

Thus, the spheroidal distance is given by

$$S = D + C_S + C_M + C_R \tag{6.6}$$

### 6.1.3 *Reduction to the Projection*

The scale factor of the projection (section 5.4.1) is applied to the spheroidal distance, $S$, given by (6.6). Therefore, using (5.28) the projection distance, $s$, is given by

$$s = KS = K(D + C_S + C_M + C_R) \tag{6.7}$$

## 6.2 Mechanical Linear Measurement

Mechanical linear measurement was once used extensively, but electromagnetic distance measurement (EDM) has to a large extent replaced it for distances of 100 m or so and upwards. For distances less than about 100 m, optical distance measurement (ODM) is often more convenient (particularly for detail surveys), but mechanical methods are usually cheaper and are often more efficient when high accuracy over short distances is necessary.

The simplest mathematical model is described in section 1.3; $R_1$ and $R_2$ are the readings at the terminals of the line and the distance can be denoted by $M = (R_1 \sim R_2)$. This is the case when the distance to be measured is less than the length of the tape. In general, the line to be measured is divided into $n$ sections or bays, each less than the tape length, and the distance can be denoted by

$$M = \sum_{i=1}^{n} (R_1 \sim R_2)_i \tag{6.8}$$

Often, in practice (for a 20 m tape, for example), $R_1 = 0$ and $R_2 = 20$ m so that

$$M = (n - 1).20 + (R_1 \sim R_2)_n \tag{6.9}$$

where $(n - 1)$ is the number of 'whole tape lengths'.

The following three sections give 'correction' terms for standardisation and temperature $(C_T)$, tension $(C_P)$ and catenary $(C_C)$, so that the refined mathematical model for $D$, the rectilinear distance between the terminals, is

$$D = M + C_T + C_P + C_C \qquad (6.10)$$

Further refinement according to sections 6.1.1, 6.1.2 and 6.1.3 may then be necessary.

### 6.2.1 *Standardisation and Temperature*

Owing to the fact that the graduations on a tape are not a true representation of multiples of the standard unit of length, use of (6.8) for $D$ will give a systematic error which can be reduced to insignificant size by a correction term. Suppose comparison is made with a higher standard of length (i.e. one which sufficiently represents a given multiple of the standard unit), giving the following results:

temperature at standardisation $= T_s$
length of the tape between terminal marks $= L_s$ (from standardisation)
nominal distance between terminal marks $= L_n (= R_1 \sim R_2)$
coefficient of thermal linear expansion $= \alpha$

The tape is then its nominal length $(L_n)$ when the temperature is $T_n$ given by

$$L_s - L_n = L_n \alpha (T_s - T_n) \qquad (6.11)$$

If the temperature during the field measurement of a bay is $T_f$, then the length of the tape is $L_f$ given by

$$L_f - L_n = L_n \alpha (T_f - T_n) \qquad (6.12)$$

and the correction term $(C_T)$ for a 'whole tape length' is given by $L_f = L_n + C_T$, where

$$C_T = L_n \alpha (T_f - T_n) \qquad (6.13)$$

and $T_n$ is calculated from (6.11).

Values of $\alpha$ for steel are around $1 \cdot 5 \times 10^{-5}$ per degC for normal tapes and around $8 \cdot 5 \times 10^{-7}$ per degC for invar tapes.

Suppose a line 100 m long is to be measured so that any error from the temperature correction term is to be less than 1 mm. If it is assumed that the only error is in the measurement of $(T_f - T_n)$ in (6.13), then the maximum tolerable error $(\delta T)$ in $(T_f - T_n)$ is, from

(1.5) and (6.13), given by $\delta C_T = L_n \alpha \, \delta T$; therefore, if $\alpha = 1 \cdot 5 \times 10^{-5}$ per degC,

$$\delta T = (1 \times 10^{-3})/(100 \times 1 \cdot 5 \times 10^{-5}) \simeq 0 \cdot 7 \text{ degC}$$

### 6.2.2 Applied Tension

The tension applied to the tape in the field $(P_f)$ may differ from that applied during the standardisation $(P_s)$. There will therefore be a correction $(C_P)$ given by

$$C_P = L_n(P_f - P_s)/AE \qquad (6.14)$$

where $L_n = (R_1 \sim R_2)$, $A$ is the cross-sectional area of the tape and $E$ is the modulus of elasticity (around $2 \cdot 0 \times 10^{11} \text{ N/m}^2$ for steel tapes).

Suppose a line 100 m long is to be measured so that any error from the tension term is to be less than 1 mm. If it is assumed that the only error is in the measurement of $P_f$ in (6.14) then the maximum tolerable error $(\delta P)$ in $P_f$ is, from (1.5) and (6.14), given by $\delta C_P = L_n \, \delta P/AE$. If $A = (10 \times 0 \cdot 5) \text{ mm}^2$ and $E = 2 \cdot 0 \times 10^{11} \text{ N/m}^2$ then $\delta P = (1 \times 10^{-3}) . (5 \times 10^{-6}) . (2 \times 10^{11})/100 = 10 \text{ N}$

### 6.2.3 Catenary Deformation

It is sometimes necessary to suspend the tape from the terminals of a bay and allow it to assume an equilibrium position under the forces of the applied tension and its own weight.

On the assumption that the distribution of the mass of the tape is uniform, it can be shown (see [1], for example) that the equation of the curve taken up by the tape is

$$y = c \cosh (x/c) \qquad (6.15)$$

which is the equation of the catenary.

In Fig 6.4, A and B are the terminals (assumed to be at the same level for the moment). The arc length AB is $(R_1 \sim R_2) = L_n$, say,

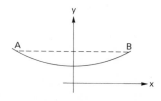

Fig 6.4

and the chord length AB is $L$, say. Then, if the terminals A and B are the same level,

$$L_n - L = w^2 L_n^3/24P^2$$

where $w$ is the weight per unit length of the tape and $P$ is the applied tension. Thus, the correction to be applied to $L_n$ to give $L$ is

$$C_C = -w^2 L_n^3/24P^2 \qquad (6.16)$$

If $\theta$ is the vertical angle between A and B, the correction becomes $C_C = -w^2 L_n^3 \cos^2 \theta/24P^2$.

Suppose a bay 50 m long is to be measured so that any error from the catenary term is to be less than 1 mm. If it is assumed that the only error is in 'weighing' the tape to obtain $w$, then the maximum tolerable error $(\delta W)$ in $W$, the weight of the tape, is, from (1.5) and (6.16), given by

$$\delta C_C = \frac{\partial}{\partial W} \left( \frac{W^2}{L_n^2} \cdot \frac{L_n^3}{24P^2} \right) \delta W = \frac{2WL_n}{24P^2} \cdot \delta W$$

If $L_n = 50$ m, $P = 40$ N and $W = 10$ N, then

$$\delta W = (1 \times 10^{-3}).(24).(40^2)/(20).(50) \simeq 0.04 \text{ N}$$

## 6.3 Electromagnetic Distance Measurement (EDM)

The tendency in modern instrumentation for EDM is towards a digital read-out, where a 'measured' value of the length of a line is obtained. The ways in which this value is arrived at are discussed in [12]. It is not proposed here to investigate how this measurement $(M)$ is obtained, but only to consider what modifications are necessary so that the measurement approaches the value for $D$, the straight line distance between the terminals of the line. There are two factors to consider: the velocity of propagation of the signal and the shape of the path traversed by the signal. These factors lead to two 'correction' terms, $C_V$ and $C_A$ respectively, which are applied to $M$ to give a value for the rectilinear distance $(D)$ between the terminals of the line:

$$D = M + C_V + C_A \qquad (6.17)$$

Examples of expressions for $C_V$ and $C_A$ are given in (6.25) and (6.42), respectively. Further modification of $D$ according to sections 6.1.1, 6.1.2 and 6.1.3 will probably be necessary.

### 6.3.1 *The Velocity of Propagation*

If $t$ is half the transit time taken by an electromagnetic signal to traverse the distance from a transmitter to a reflector and back again, then the distance between the transmitter and the reflector is $M_1 = vt$ where $v$ is the velocity of the signal along the path. Ways in which $t$ is effectively measured are described in [12]. In most EDM instruments which give a digital read-out representing $M$, the measured distance, a value $(u)$ for the velocity is assumed which may be close enough to the actual velocity $(v)$. However, it is often necessary to modify $M = ut$ to give $M_1 = vt$. This correction $(C_V)$ is given by

$$C_V = M_1 - M = M.\frac{v}{u} - M = M\left(\frac{v}{u} - 1\right) \qquad (6.18)$$

It is therefore necessary to estimate the velocity of propagation $(v)$ at the time of the measurement. This is done by measuring the atmospheric conditions.

For a sine wave emitted with frequency $f$ and travelling through a vacuum with velocity $c$, the wavelength of the radiation is $\lambda = c/f$. The accepted (1972) value of $c$ is $299\ 792 \cdot 5 \pm 0 \cdot 4$ km/s. The velocity of the signal through the atmosphere $(v)$ is related to the *refractive index*, $n$, by the equation $n = c/v$, and $n$ itself depends upon the frequency of the signal and the atmospheric conditions. Cauchy's equation

$$n = A + \frac{B}{\lambda^2} + \frac{C}{\lambda^4} \qquad (6.19)$$

describes the *dispersion*, i.e. the relation between the refractive index and the wavelength $\lambda = c/f$ for normal dispersion. $A$, $B$ and $C$ are constants for given atmospheric conditions. For example, at 0 °C and $1013 \cdot 25$ mbar (760 mm Hg) pressure, dry air with $0 \cdot 03\%$ $CO_2$ content has refractive index $n_s$ given by

$$(n_s - 1).10^6 = 287 \cdot 604 + \frac{1 \cdot 6288}{\lambda^2} + \frac{0 \cdot 0136}{\lambda^4} \qquad (6.20)$$

where $\lambda$ is the wavelength $(= c/f)$ in μm [10].

The refractive index under other atmospheric conditions also depends upon the wavelength of the radiation. For example, for instruments using radiation in or near the visible spectrum, the refractive index is given by [10]:

$$(n_t - 1) = \frac{(n_s - 1)}{\alpha T}.\frac{P}{1013 \cdot 25} - \frac{11 \cdot 27E}{T}.10^{-6} \qquad (6.21a)$$

where $n_s$ is from (6.20), $P$ is the total atmospheric pressure in mbar, $T$ is the temperature in °K, $\alpha$ is the coefficient of thermal expansion of air (approximately $1/273 \cdot 16$) and $E$ is the partial pressure of the water vapour content in mbar. For microwaves, the formulae of Essen and Froome [14], quoted in [12], are applicable, and an approximate (better than 1 in $10^6$) formula is

$$(n_t - 1) . 10^6 = 77 \cdot 624 . \frac{(P - E)}{T} + \frac{64 \cdot 7}{T} \left( 1 + \frac{5748}{T} \right) . E$$

$$(6.21b)$$

where the units are the same as for (6.21a). If $E = 0$, $P = 1013 \cdot 25$ mbar and $T = 273$ °K, then $(n - 1) . 10^6 \simeq 288$, and this is the value given by (6.20), since $\lambda$ is of the order of 10 mm ($10^4$ μm) and the terms in $\lambda^2$ and $\lambda^4$ in (6.20) are negligible.

So far, in this section, it has been assumed that the signal is a sine wave; however, in practice it is a modulated sine wave where the modulation frequency is the one which is important for distance measurement. Thus the *group velocity* [12] is the velocity to be used and this is given by the *group refractive index*:

$$n_g = n - \lambda . \partial n / \partial \lambda \qquad (6.22)$$

Substitution of $n$ and $(\partial n / \partial \lambda)$ from (6.19) gives

$$n_g = A + \frac{3B}{\lambda^2} + \frac{5C}{\lambda^4} \qquad (6.23)$$

and, under the atmospheric conditions relevant to (6.20),

$$(n_{gs} - 1) . 10^6 = 287 \cdot 604 + \frac{4 \cdot 8864}{\lambda^2} + \frac{0 \cdot 0680}{\lambda^4} \qquad (6.24)$$

For visible light, the group refractive index ($n_{gt}$) under any other atmospheric conditions is given by substituting $n_{gs}$ for $n_s$ in (6.21a). For microwaves, since $(\partial n / \partial \lambda)$ is negligible, the group velocity is the same as the phase velocity.

Thus, the expression (6.18) for $C_V$ becomes

$$C_V = M \left( \frac{n_a}{n_{gt}} - 1 \right) \qquad (6.25)$$

where $M$ is the measured value (based on an assumed refractive index $n_a$) and $n_{gt}$ is the refractive index, calculated from (6.21a) and (6.24) in the case of visible and near-visible waves, and from equations similar to (6.21b) in the case of microwaves. The use of (6.21a) and (6.21b) necessitates obtaining values of $P$, $T$ and $E$, usually by taking readings at one end of the line in the case of visible and near-visible waves, and the average of readings at each end of the line in the case of micro-waves. In such practical cases, the assumption that atmospheric con-ditions measured a few feet above ground level are characteristic of the line as a whole introduces systematic error which is difficult to reduce, and can amount to several parts per million. A small addi-tional term can be used to reduce some of the error and it is shown in [1] that this term is $-M^3 k(1 - 2k)/6R_a^2$, where $k$ is the coefficient of refraction defined in (6.39).

As a numerical example, consider the Tellurometer MA 100 which uses a carrier frequency with a wavelength in vacuo of 0·93 μm. The group refractive index of dry air at 0 °C and 1013·25 mbar pressure is, from (6.24), given by

$$(n_{gs} - 1).10^6 = 287\cdot604 + \frac{4\cdot8864}{(0\cdot93)^2} + \frac{0\cdot0680}{(0\cdot93)^4} \simeq 293\cdot34$$

Substitution of this value in (6.21a) gives the group refractive index $(n_{gt})$ from

$$(n_{gt} - 1).10^6 = \frac{293\cdot34}{\alpha T} \cdot \frac{P}{1013\cdot25} - 11\cdot27 \cdot \frac{E}{T}$$

so $$(n_{gt} - 1).10^6 = 79\cdot08 \cdot \frac{P}{T} - 11\cdot27 \cdot \frac{E}{T} \qquad (6.26)$$

which is quoted by [17], but with pressures in mm Hg. The velocity to be used in the correction term (6.18) is $v = c/n_{gt}$. The MA 100 uses a refractive index of $n_a = 1\cdot000\ 274$ to obtain the digital display of the measured value $M$, so for this instrument (6.25) becomes

$$C_V = M \left( \frac{1\cdot000\ 274}{n_{gt}} - 1 \right)$$

where $n_{gt}$ is obtained from (6.26).

Since, in practice, barometer and thermometer readings are made in order to calculate $n_{gt}$ and, hence, $C_V$, it is useful to consider what are the effects of errors in temperature and pressure readings on the correction term.

1. Suppose the only error is in the measurement of $P$. Then, for 0·93 μm wavelength radiation, differentiation of (6.26) gives

$$10^6 . \delta n_{gt} = 79·08 . \delta P/T$$

If $\delta n_{gt}$ (and hence the distance) is to have a maximum error of 1 in $10^6$, the largest error in $P$ which can be tolerated is (at $T = 283$ °K), $\delta P \simeq 3·6$ mbar.

2. Suppose the only error is in the measurement of $T$. Then, for 0·93 μm wavelength radiation, differentiation of (6.26) gives

$$10^6 . \delta n_{gt} = -79·08 P . \frac{\delta T}{T^2} + 11·27 E . \frac{\delta T}{T^2} \qquad (6.27)$$

and taking $\delta n_{gt}$ as 1 in $10^6$ as before, at $T = 283$ °K, $P = 1013$ mbar, $\delta T \simeq -1$ degC, the second term being insignificant.

3. Suppose the only error is in the measurement of $E$. Then, for 0·93 μm wavelength radiation, differentiation of (6.26) gives $10^6 . \delta n_{gt} = -11·27 . \delta E/T$ and, under the above conditions, $\delta E \simeq -25$ mbar. $E$ is given by [17] as

$$E = 10^{[9·6 - (2450/T')]} - 6·7(T - T')P . 10^{-4} \text{ mm Hg}$$

where $T'$ is the wet-bulb temperature in °K. An error of 25 mbar in $E$ is caused by a 4 degC error in the wet-bulb temperature, if $P = 1013$ mbar and $\delta T = 4$ degC.

For microwaves, similar analysis of (6.21b) indicates that for 1 in $10^6$ accuracy, $\delta T \simeq 1$ degC, $\delta P \simeq 3.6$ mbar, $\delta E \simeq 0·2$ mbar and $\delta T' \simeq 0·1$ degC, and the water vapour content is thus far more important.

### 6.3.2 Curvature of the Path

Fig 6.5 shows a vertical section along the ray path which passes through concentric layers of atmosphere with centre at O, the centre of the (spherical) earth.

Considering the refraction at B between layers with refractive index $n_1$ and $n_2$ respectively:

$$n_1 \sin i = n_2 \sin r \qquad (6.28)$$

But from triangle OBA:

$$\sin i = \frac{OA}{OB} \sin p \qquad (6.29)$$

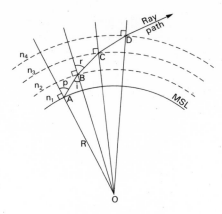

Fig 6.5

Therefore, substituting for sin $i$ in (6.28):

$$n_1 . OA . \sin p = n_2 . OB . \sin r \qquad (6.30)$$

Therefore, along the ray path

$$(R + h)n \sin \phi = \text{constant}$$

where $R$ is the radius of the earth, $h$ is the height above MSL, $n$ is the refractive index and $\phi$ is the zenith angle. Putting $\alpha = 90 - \phi$,

$$(R + h)n \cos \alpha = \text{constant} \qquad (6.31)$$

In Fig 6.6, PQ is a small element of the ray path of length $\delta s$ (greatly exaggerated) with P at $(R + h)$ above MSL and Q at $(R + h + \delta h)$

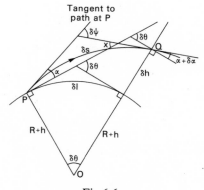

Fig 6.6

above MSL. The curvature is defined by $(1/\sigma) = (d\psi/ds)$, $\delta\psi$ being the angle between the tangents to the path at P and Q. It can be seen by considering angle $x$ in Fig 6.6 that $\delta\psi = (\delta\theta - \delta\alpha)$ so that the curvature can be written as

$$\frac{1}{\sigma} \simeq \frac{\delta\theta - \delta\alpha}{\delta s} \tag{6.32}$$

But $\qquad \delta\theta = (\delta l)/(R + h) \quad \text{and} \quad \delta s \simeq \delta l . \sec \alpha$

Therefore $\qquad \dfrac{1}{\sigma} \simeq \dfrac{[(\delta l)/(R + h)] - \delta\alpha}{\delta l . \sec \alpha}$

$$\simeq \frac{\cos \alpha}{R + h} - \frac{\delta\alpha}{\delta l} \cos \alpha$$

But $\delta l \simeq \delta h \cot \alpha$, so that

$$\frac{1}{\sigma} \simeq \frac{\cos \alpha}{R + h} - \frac{\delta\alpha . \cos \alpha}{\delta h . \cot \alpha}$$

Therefore $\qquad \dfrac{1}{\sigma} \simeq \dfrac{\cos \alpha}{R + h} - \dfrac{d\alpha}{dh} . \sin \alpha \tag{6.33}$

Logarithmic differentiation of (6.31) gives

$$\frac{dh}{R + h} + \frac{dn}{n} - \frac{\sin \alpha}{\cos \alpha} . d\alpha = 0$$

so that $\qquad d\alpha = \left( \dfrac{dh}{R + h} + \dfrac{dn}{n} \right) . \cot \alpha$

substitution of which in (6.33) gives

$$\frac{1}{\sigma} \simeq -\frac{1}{n} \frac{dn}{dh} . \cos \alpha \tag{6.34}$$

The gradient $dn/dh$ for light waves can be found by differentiation of (6.21a) which gives

$$\frac{dn_{gt}}{dh} = \frac{(n_{gs} - 1)}{\alpha T} . \frac{1}{1013 \cdot 25} \left( \frac{dP}{dh} - \frac{dT}{dh} . \frac{P}{T} \right)$$

since the term including $E$ is insignificant. Thus

$$\frac{dn_{gt}}{dh} = \frac{0 \cdot 270(n_{gs} - 1)}{T} \left( \frac{dP}{dh} - \frac{dT}{dh} . \frac{P}{T} \right) \tag{6.35}$$

where $n_{gs}$ is given by (6.24), $T$ is in $°K$, $(dP/dh)$ is in mbar/m, $(dT/dh)$ is in degC/m and $P$ is in mbar.

Substitution of this expression for the refractive index gradient in (6.34) gives the general expression for the curvature of a light path in terms of the atmospheric parameters $P$, $T$, $(dP/dh)$ and $(dT/dh)$:

$$\frac{1}{\sigma} \simeq \frac{0{\cdot}270(n_{gs} - 1)\cos\alpha}{n_{gs}.T}\left(\frac{P}{T}.\frac{dT}{dh} - \frac{dP}{dh}\right)\text{m}^{-1}$$

In practice, to determine correction terms, both $\cos\alpha$ and $n_{gs}$ can be taken as unity and $(n_{gs} - 1)$ as $0{\cdot}000\ 293$ for visible light, so that

$$\frac{1}{\sigma} \simeq \frac{0{\cdot}790}{10^4.T}\left(\frac{P}{T}.\frac{dT}{dh} - \frac{dP}{dh}\right)\text{m}^{-1} \tag{6.36}$$

The curvature therefore depends upon the vertical temperature and pressure gradients. [11] gives average daytime values for these as

$$\frac{dP}{dh} \simeq -0{\cdot}0342(P/T)\ \text{mbar/m} \tag{6.37}$$

and $\qquad dT/dh \quad$ between $-0{\cdot}01$ and $-0{\cdot}0055$ degC/m $\qquad$ (6.38)

The latter (temperature) gradient is by far the most variable in practice. For example, at night, *temperature inversion* is normal and $(dT/dh)$ is positive. For methods which enable $(dT/dh)$ to be estimated, see [3]. With these values and $P = 1000$ mbar and $T = 300\,°K$ as an example, (6.36) gives $\sigma \simeq 43 \times 10^6$ m. It is generally assumed for the purpose of calculating corrections that the curvature of path is constant over distances involved in EDM with wavelengths in or near the visible spectrum. The ratio of the radius of the spheroid $(R_\alpha)$ to $2\sigma$ is often referred to as the *coefficient of refraction* $(k)$:

$$k = (R_\alpha/2\sigma) \tag{6.39}$$

Sometimes, however, the coefficient of refraction is defined as $(R_\alpha/\sigma)$. The definition of $k$ in (6.39) is used again in sections 6.5.1 and 6.5.2.

For visible light, it is shown above that $\sigma \simeq 43 \times 10^6$ m and, taking $R_\alpha = 6.36 \times 10^6$ m, it is seen that a typical value for the coefficient of refraction $k = (R_\alpha/2\sigma)$ is $0{\cdot}07$.

In the case of microwave EDM, differentiation of Essen and Froome's formula (6.21b) to give $(dn/dh)$, and subsequent simplification using $(dP/dh)$ and $(dT/dh)$ from (6.37) and (6.38) respectively, gives

$$\frac{dn}{dh} = -\frac{2 \cdot 22}{10^6} \cdot \frac{P}{T^2} + 0 \cdot 0042 \cdot \frac{E}{T^3} + \frac{0 \cdot 38}{T^2} \cdot \frac{dE}{dh} \tag{6.40}$$

so that substitution in (6.34) with $n \simeq \cos \alpha \simeq 1$ gives the curvature of the path for microwaves:

$$\frac{1}{\sigma} \simeq \frac{2 \cdot 2}{10^6} \cdot \frac{P}{T^2} - 0 \cdot 0042 \frac{E}{T^3} - \frac{0 \cdot 38}{T^2} \cdot \frac{dE}{dh} \tag{6.41}$$

If $P = 1000$ mbar, $T = 300\ °\mathrm{K}$, $E = 20$ mbar and $dE/dh = -0 \cdot 007$ mbar/m, then $\sigma \simeq 20 \times 10^6$ m and $k = 0 \cdot 16$. Various atmospheric conditions are described in [11] under which $k$ is seen to vary between 1 and $0 \cdot 1$; a value of $0 \cdot 125$ can be taken as standard.

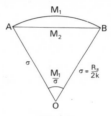

Fig 6.7

In Fig 6.7, $M_1$ is the arc length from (6.18) and (6.25) and $M_2$ is the chord length. The radius $\sigma$ of the circular path is given by (6.39) as $(R_\alpha/2k)$, where $k \simeq 0 \cdot 07$ for light and $k \simeq 0 \cdot 125$ for microwaves. Thus the correction is $C_A = (M_2 - M_1)$.

But
$$M_2 = 2\sigma \sin (M_1/2\sigma)$$

$$= 2\sigma \left( \frac{M_1}{2\sigma} - \frac{M_1^3}{48\sigma^3} + \cdots \right)$$

Therefore
$$M_2 \simeq M_1 - \frac{M_1^3}{24\sigma^2}$$

or
$$(M_2 - M_1) = C_A \simeq -\frac{M_1^3 k^2}{6R_\alpha^2} \tag{6.42}$$

Thus, by substitution in (6.17) from (6.42) and (6.25)

$$D = M + M \left( \frac{n_a}{n_{gt}} - 1 \right) - \frac{M_1^3 k^2}{6R_\alpha^2}$$

and
$$D \simeq M \left( \frac{n_a}{n_{gt}} - \frac{M^2 k^2}{6 R_\alpha^2} \right)$$
(6.43)

## 6.4 Optical Distance Measurement (ODM)

Methods of ODM are described fully in [23], where it can be seen that a correction to horizontal, for example, is often made by a combination of optical and mechanical reduction components in the instruments. The effects of errors in reduction systems are also discussed.

## 6.5 Vertical Angle Measurement

Vertical angles are generally measured for one or both of two reasons: to enable a slope distance to be reduced to the horizontal according to (6.2a) or to determine the height difference between the terminals of a line (trig. heighting, see section 4.3.2). The suitable mathematical model will be different in each case since the required accuracy is different. The effect of atmospheric refraction on the vertical angle and the effect of earth curvature on the calculated height difference must be considered.

### 6.5.1 *The Effect of Atmospheric Refraction on the Vertical Angle*

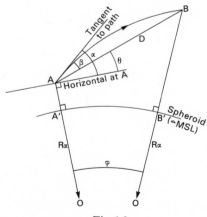

Fig 6.8

In Fig 6.8, $\theta$ is the vertical angle at A between the horizontal at A and the chord AB (of length $D$). However, owing to atmospheric refraction, the observed angle at A is $\alpha$, between the tangent to the path at A and the horizontal. Thus there is a correction $\beta$ to the observed value $\alpha$ to give $\theta$: $\theta = \alpha - \beta$. If $\sigma$ is the radius of the path (which is assumed constant) then arc AB $= 2\sigma\beta$ and arc $A'B' = R_\alpha\phi$. Since AB $\simeq A'B'$, $2\sigma\beta \simeq R_\alpha\phi$ and, therefore,

$$\beta \simeq R_\alpha\phi/2\sigma \simeq D/2\sigma \qquad (6.44)$$

But the coefficient of refraction defined by (6.39) is $k = (R_\alpha/2\sigma)$, therefore, substituting for $\sigma$ in (6.44),

$$\beta \simeq kD/R_\alpha \text{ radians} \qquad (6.45)$$

### 6.5.2 *The Effect of Earth Curvature on the Calculated Height Difference*

If the earth is assumed to be flat, then the height difference between A and B in Fig 6.1 is BC and is equal to D sin $\theta$, where $\theta$ is the observed angle corrected for refraction by subtracting $\beta$, which is given by (6.45). However, when the effect of earth curvature is considered, the height difference between A and B becomes

$$\text{BE} = D \sin (\theta + \gamma) \sec (\phi/2)$$

from triangle ABE in Fig 6.1. Generally, without loss of accuracy, sec $(\phi/2)$ can be put equal to unity, so that the effect of earth curvature can be allowed for by the addition of a correction $\gamma$ to $\theta$, where

$$\gamma \simeq \phi/2 = \text{arc } A'B'/2R_\alpha = S/2R_\alpha \simeq D/2R_\alpha \qquad (6.46)$$

where $S$ is from (6.6).

It is often convenient to combine the refraction and curvature corrections so that the measured angle $\alpha$ becomes $\alpha' = \alpha - \beta + \gamma$ where $\beta$ is given by (6.45) and $\gamma$ is from (6.46). Thus

$$\alpha' \simeq \alpha - \frac{kD}{R_\alpha} + \frac{D}{2R_\alpha} = \alpha + \frac{D}{2R_\alpha}(1 - 2k) \qquad (6.47)$$

If $D = 1$ km and $k = 0.07$ (from section 6.3.2) then $D(1 - 2k)/2R_\alpha$ is approximately $14''$. Thus the effects of curvature and refraction on trig. heights can be effectively reduced by using $\Delta h = D \sin \alpha'$, where $\alpha'$ is given by (6.47).

## 6.6 Horizontal Angle Measurement

Measurement of a horizontal angle entails the observation of two directions, and the 'measured' angle is the difference between the measured directions. The effect of atmospheric refraction on a measured direction is considered in section 6.6.1. Since an angle is measured between three points on (or near) a spheroidal surface, the sum of the angles of the triangle should be greater than 180°. This is considered in section 6.6.2.

### 6.6.1 *Lateral Refraction*
The atmospheric refraction of light in a vertical plane has been considered in section 6.3.2 and its effect on vertical angle measurement is described in section 6.5.1. For any horizontal direction measured with a theodolite, there will be some refraction in a horizontal plane, although this *lateral refraction* is very much less than the refraction in a vertical plane and is often insignificant. Its magnitude depends upon the horizontal temperature and pressure gradients. The latter can generally be ignored and the former will be greater where a ray passes close to the shoulder of a hill, the edge of a building or between two different types of ground surface. The curvature in a horizontal plane is, from (6.36), given by

$$\frac{1}{\sigma} \simeq \frac{0.790}{10^4} \cdot \frac{P}{T^2} \cdot \frac{dT}{dy} \, m^{-1} \tag{6.48}$$

where $(dT/dy)$ is the horizontal temperature gradient in °C/m and the pressure $P$ is in mbar. It is shown in [11] that the effects of lateral refraction are of the order of $1''$ under specified conditions. Generally it is not possible to calculate useful correction terms owing to the difficulties in determining $(dT/dy)$ in the field. However, for precise surveys in enclosed spaces (workshops, tunnels, etc.) a significant horizontal temperature gradient could be allowed for, since conditions are likely to be stable enough for measurement of $(dT/dy)$.

The effects of this systematic error can be reduced in the field, firstly by avoiding grazing rays and secondly by averaging observations made at times when $(dT/dy)$ is of opposite sign (e.g. in the day and at night). This latter method is an example of the application of the principle of reversal (section 1.3.1).

### 6.6.2 *Spherical Excess*
For a spherical triangle, the *spherical excess*, or the amount by which

the angles of the triangle exceed 180°, is given by

$$\epsilon = A/R^2 \text{ rad} \qquad (6.49)$$

where $A$ is the area of the triangle and $R$ is the radius of the sphere. In the case of a spheroidal triangle, the excess is $\epsilon = A/\rho v$ where $(\rho v)$ is taken for the latitude of the centroid of the triangle.

*Legendre's theorem* (verified in [1], for example) states that each angle of a plane triangle, whose sides are the same as the sides of a spheroidal triangle, is equal to the corresponding angle of the spheroidal triangle, less $\epsilon/3$. The same applies to a spheroidal triangle, except when the sides are very long (over 100 km) and the accuracy required is very great (less than 0"·1).

The main application of Legendre's theorem is in the solution of spheroidal triangles. In Fig 6.9, ABC is a spheroidal triangle with spherical excess $\epsilon$, spheroidal distances $a$, $b$ and $c$ and spheroidal angles A, B and C.

Fig 6.9

Legendre's theorem means that rules of plane trigonometry can be applied, provided the angles are each reduced by $\epsilon/3$. Thus for example,

$$\frac{a}{\sin (A - \epsilon/3)} = \frac{b}{\sin (B - \epsilon/3)} = \frac{c}{\sin (C - \epsilon/3)} \qquad (6.50)$$

where $a$, $b$ and $c$ can be used to compute the area of the triangle without significant error.

# References

1. ALLAN, A. L., HOLLWEY, J. R. and MAYNES, J. H. B. *Practical Field Surveying and Computations*. Heinemann. 1968

2. ANDERTON, P. and BIGG, P. H. *Changing to the Metric System*. HMSO, London. 1967

3. ANGUS LEPPAN, P. V. 'A study of refraction in the lower atmosphere.' *Empire Survey Review*. 120, 121, 122. 1963

4. *Constants, Formulae and Methods Used in Transverse Mercator Projection*. HMSO, London. 1950

5. *Projection Tables for the Transverse Mercator Projection of Great Britain*. HMSO, London. 1950

6. *Universal Transverse Mercator Grid Tables*. Directorate of Military Survey, London. 1958

7. Technical Manual TM–5–241–3, US Army Map Service, Washington, DC

8. ASHKENAZI, V. 'Solution and error analysis in large geodetic networks. Direct methods.' *Survey Review*. 146. pp 166–173. 1967. 147. pp 194–206. 1968

9. ASHKENAZI, V. 'Adjustment of control networks for precise engineering surveys.' *Chartered Surveyor*. 102. 7. pp 314–320. 1970

10. BARRELL, H. and SEARS, J. E. 'The refraction and dispersion of air for the visible spectrum.' *Phil. Trans. R. Soc.* A238. 1939

11. BOMFORD, G. *Geodesy*. Clarendon. 1971

12. BURNSIDE, C. D. *Electromagnetic Distance Measurement*. Crosby Lockwood. 1971

13. CRAMER, H. *Random Variables and Probability Distributions*. pp 19–20. Cambridge. 1937

14. ESSEN, L. and FROOME, K. D. 'The refractive index of air for radio-waves and microwaves.' *Proc. Phys. Soc.* B54. pp 862–873. 1951

15. GAUSS, C. F. *Theoria Motus Corporum Coelestium*. Lib 2, Sec. 3. 1809

16. HALD, A. *Statistical Theory with Engineering Applications*. Wiley. 1952

17. HÖLSCHER, H. D. *A Short-range Highly Accurate Electro-optical Distance Measuring Equipment*. National Institute for Telecommunications Research, C.S.I.R. S. Africa. Undated

18. LEE, L. P. 'The Transverse Mercator projection of the spheroid.' *Empire Survey Review*. 38. pp 142–152. 1945

19. LEGENDRE, A.-M. *Nouvelles Méthodes pour la Détermination des Orbites des Comètes*. Paris. 1806

20. LOGAN, W. R. 'The rejection of outlying observations.' *Empire Survey Review*. 97. pp 133–137. 1955

21. PEARSON, E. S. and CHANDRA SEKA, C. 'The efficiency of statistical tools and a criterion for the rejection of outlying observations.' *Biometrika*. 28. 1936

22. RAINSFORD, H. F. *Survey Adjustments and Least Squares*. Constable. 1957.

23. SMITH, J. R. *Optical Distance Measurement*. Crosby Lockwood. 1970

24. THOMPSON, E. H. *Introduction to the Algebra of Matrices with Some Applications*. Hilger. 1969

25. WHITTAKER, E. and ROBINSON, G. *The Calculus of Observations*. Blackie. 1944

# Index